One Voice

Pacifist Writings from the Second World War

Humiliation with Honour

Seed of Chaos

VERA BRITTAIN

With a Foreword by Shirley Williams
and an Introduction by Y. Aleksandra Bennett

Continuum

The Tower Building
11 York Road
London SE1 7NX

15 East 26th Street
New York
NY 10010

www.continuumbooks.com

Foreword copyright © Shirley Williams, 2005
Introduction copyright © Y. Aleksandra Bennett, 2005
Vera Brittain material copyright © Mark Bostridge and Rebecca Williams, Literary Executors of the Vera Brittain Estate, 2005

All rights reserved. No part of this publication may be reproduced or transmitted in any form or by any means, electronic or mechanical, including photocopying, recording or any information storage or retrieval system, without prior permission from the publishers.

First published 2005

British Library Cataloguing-in-Publication Data
A catalogue record for this book is available from the British Library.

ISBN 0–8264–8534–0

Typeset by RefineCatch Limited, Bungay, Suffolk
Printed and bound by MPG Books Ltd, Bodmin, Cornwall

Pacifism is nothing other than a belief in the ultimate transcendence of love over power. This belief comes from an inward assurance. It is untouched by logic and beyond argument – though there are many arguments both for and against it. And each person's assurance is individual; his inspiration cannot arise from another's reasons, nor can its authority be quenched by another's scepticism.

Vera Brittain, 1942

Contents

Foreword

Shirley Williams

The First World War was the spectre, the unseen presence, in my mother's writings, at her moments of decision, in almost everything she did. The contrast between the idealism of the young men (and some young women) who went to war in 1914, and the terrible toll of deaths and injuries to no great end, haunted the survivors of her generation. *One Voice* has to be read against that background.

The years between the two great World Wars were only an interval in the long civil war of Europe. They were years in which people strove to return to a stable life, or at least to put the horrors of the war behind them. They were years in which, as Aleksandra Bennett points out, my mother achieved renown for her autobiography of the First World War, *Testament of Youth*. She enjoyed the celebrity that came her way, the Foyles lunches with other well-known writers, the invitations to lecture, her lionization on the American tours she undertook in the 1930s. She could feel, briefly, that she had accomplished her self-imposed mission, that the four young men she had loved and lost would not be forgotten, and that war would be seen for what it was: cruel, bitter, obscene.

But she learned, as we all do, that such missions are never finally accomplished: they go on demanding our time, our energy, our commitment, for as long as we live. My mother's achievement in the seven years after the publication of *Testament of Youth* with other books (like *Honourable Estate* and *Testament of Friendship*), brought her no tranquillity. The invasion of Abyssinia by Italy, the annexation of the Rhineland by Hitler, and the failure of the League of Nations to intervene, together with the false dawn of Munich, were warning signs of a Europe again gearing up for war.

After toying briefly with the idea of sitting out the war as a writer and mother in some rural English village, my mother confronted the probability that this would be no skirmish, no phoney war, but another worldwide conflict. At first she shared the commonly held view that the war would be short. When my brother John and I were evacuated to the United States in July 1940, my parents spoke of six months or at most a year. My mother still fondly hoped she would be able to visit us on her lecture tours. Meanwhile, she continued to denounce the war and to urge a peaceful compromise, believing that nothing could be worse than another World War.

My father was already involved in a very different cause, that of trying to persuade the United States to enter the war on the side of the European allies, Britain and France. He was not a pacifist, and did not believe it was possible to reach any lasting settlement with Hitler's Third Reich. A friend of Wendell Willkie, the Republican Presidential candidate in 1940, he travelled on the campaign train that autumn as an advisor on foreign policy.

After the election, as the Battle of the Atlantic threatened the supply routes to Britain, he lobbied for the bases for ships deal, and for lend-lease, the first tentative steps in an American commitment to the war. Torpedoed on one Atlantic crossing, he arrived at Paddington station in salt-stained silk pyjamas and a fisherman's jersey, having been rescued by an Irish fishing boat.

The long separation between my parents and us, their children, was formalized when my mother was refused a visa to visit the United States, the scene of her earlier triumphs. Coupled with attacks on her for her pacifism, and rejection by many she had thought of as friends, she felt as if she were entering a long dark tunnel.

But paradoxically, she found in the war itself a new way of pursuing her lifelong mission. She began the series of *Letters to Peace-Lovers*, sent out regularly to a network of devoted friends and followers. Sometimes the letters were written and typed as air raids raged around her and her devoted young assistant, Winfred Eden-Green. She became a firewatcher, and later travelled the country raising funds for the Peace Pledge Union's Food Relief Campaign.

Humiliation with Honour combined the lessons of the spiritual ordeal my mother experienced in the war years: the humiliation, and then the dawning of a realization that this was the way to a deeper understanding of the human condition and of the radical teaching of the Christian Gospels, with the profound need to communicate with the teenage son she loved. John Edward, born in 1927, had been named after her brother Edward, killed in battle in Italy in the last year of the First World War. John's deep brown hair, marked eyebrows and sensitive face resembled Edward as a child. My mother yearned for him to appreciate what she was doing and why she was doing it. Her letters have no guile; they are transparent in their honesty.

It is unlikely that John sympathized with my mother's letters. He had been sent by his American guardian, George Brett – the head of Macmillan, my mother's publisher in the United States – to a school with a strong military tradition. Minnesota, where my brother and I lived with our American foster-parents – a former Navy surgeon, later a much respected doctor, and his wife, a pacifist with a strong connection to international causes – was isolationist and deeply suspicious of foreign entanglements. After the Japanese attack on Pearl Harbour, however, the middle western state became unquestioningly enthusiastic for the war. For John, his mother's personal mission must have been something of an embarrassment.

That mission was largely ignored by the public until the summer of 1943. The massive British and American raids on German cities were part of the strategy of breaking German civilian morale. The fine distinctions between civilians and soldiers under the Geneva Conventions were not only forgotten: they were reversed. The objective became the maximum destruction of the German war machine, and then just of Germany. A large part of Germany's cultural heritage was wiped out, along with hundreds of thousands of children and others in no way engaged in warfare.

My mother hesitated before committing herself to this cause. She had already been through much obloquy for her pacifism. She had some idea of the consequences of denouncing mass bombing. Either she had been forgotten, and her intervention would go unnoticed, or she was not forgotten, and would be condemned by many who were passionately engaged in attaining victory whatever the cost.

The latter turned out to be the case, at least in the United States. From the abrasive journalist Dorothy Thompson to Franklin and Eleanor Roosevelt themselves, Vera Brittain was regarded at best as a mushy sentimentalist, at worst as an unthinking supporter of the Nazis. Bishops, writers and legislators hurried to add their voices to the condemnation. Only the Quakers, the Fellowship of Reconciliation and a handful of Christian activists, prominent among them the great American Catholic Dorothy Day, defended her.

Solace came from an unexpected quarter. In 1945, the Black List of people living in Britain to be immediately arrested after a German invasion was discovered and then published, making the front page of most British newspapers. The page of the list most editors chose to print was the one bearing the name of Winston Churchill. Luckily, it was also the page bearing the names of both my parents. No more effective riposte could have been discovered. Few of my mother's critics found their way on to the list.

Evidence that questioned the whole strategy of mass bombing was slower to come. The United States air force undertook a detailed study of the effects of mass bombing as compared to precision bombing. One of those who participated in the study was the great economist, John Kenneth Galbraith. Following the huge raids, he told me, the German government stepped up its mobilization of civilians, including married women with children. The journalist and war historian Max Hastings cast doubt on the efficacy of mass bombing in his book on Bomber Command, published in 1979. Even setting all moral considerations aside, the arguments were far from conclusive.

You cannot set aside moral considerations, my mother would have said. Two of her most passionately held themes were that wars corrupt language and thereby twist truth, and that actions sanctioned by war blunt human sympathy and human sensitivity. Because human beings, in her opinion, are basically good, the cruelties of war have to be represented in terms drained of emotion. So massive numbers of civilian deaths become collateral damage. Shooting or bayoneting injured enemies becomes cleaning up. Imprisoning and sometimes torturing uncharged suspects is detention. We are good at weasel words.

Worse still is the hardening of hearts and minds to the viciousness of war. The long struggle to build a world of law, to embody respect for human life and human rights in our international and national institutions is always damaged by war, today in Iraq, the Middle East and Africa as much as once in Europe. In that struggle, the voices of men and women of uncomfortable conscience remain indispensable. I am proud that my mother was amongst them.

Introduction

Y. Aleksandra Bennett

In what 'practical' terms should a liberal democracy confront a force that threatens to overwhelm it? This question engaged Vera Brittain (1893–1970) with singular immediacy throughout the years of the Second World War. She was a pacifist in a war against Hitler's Nazi Germany. She was also a proud and ambitious woman, who enjoyed the considerable material success and international recognition that stemmed from the publication of her remarkable autobiography, *Testament of Youth* (1933).[1] In January 1937, Canon Dick Sheppard, founder of the Peace Pledge Union (PPU), expressed delight that Brittain had agreed to join, not least because of 'what your name means to the cause'.[2] Within five years, however, her sponsorship of the PPU and her unpopular and controversial stance on the war had made Brittain an outcast. Her name was associated with Fifth Columnists, and there were allegations of both Nazi and Communist sympathies.[3] Over time these perceptions tarnished her prestige, diluted her influence, lessened her social acceptance and diminished her book sales in the United States. In short, many of the things Vera Brittain had craved and which were for her measures and outward signs of her success were replaced by official suspicion and public hostility, social ostracism and private criticism.

Brittain met these misfortunes with deep resentment. They were bitter blows, spawning frustration, anger and self-pity. But, aware that her own experience as a pacifist would have been very different if the British Home Secretary were a Nazi and she a German pacifist, she struggled against these feelings.[4] From the sidelines, some advised Brittain to recant her views, or to disassociate herself from the PPU, or at the very least to lie 'doggo'.[5] She did none of these things.

Within a week of the outbreak of the Second World War, Vera Brittain declared her opposition to war in an article entitled 'What Can We Do In War Time?'[6] She recognized that there were pacifists, like herself, who identified with the broader community and who, without compromising their pacifist principles, wanted to contribute to society. She urged them to seize every opportunity to advance the cause of peace and conciliation; to challenge the distortions of war-time propaganda; to labour for the amelioration of suffering at home and abroad; and to work for reconstruction in order to lay 'the foundation of a just and lasting peace'.[7] The article was the manifesto by which Brittain tried to live: *Humiliation with Honour* and *Seed of Chaos* were two of its expressions.[8]

Brittain had considered what her pacifism in wartime might entail. In mid-January 1939, writing to a friend, Brittain had expressed the hope that she might be able to spend any war in the country, writing and looking after her two children, or possibly working with the Quakers. Although Brittain admired the Society of Friends, she was neither a Quaker, nor was she, when she joined the PPU, a religious pacifist. She counted herself a pacifist for 'quite other than religious reasons'.[9] Indeed, she foresaw a split in the PPU between the political and religious inspirations of pacifism. She believed that it was vitally important for the political side of the movement to be kept alive in order to counterbalance those who might otherwise get out of touch with political realities.[10]

Brittain was to encounter unanticipated political realities of her own in the next few months. In the autumn of 1939, she undertook an exhausting schedule of meetings across the country, speaking in favour of an armistice and an early, negotiated peace.[11] In January 1940, she had government approval to travel to the United States to deliver a series of lectures arranged, before the war, by her American agents. She returned to Britain in early April 1940, just as the phoney war was ending. A mood of apprehension swept over the PPU in the spring of 1940, with regular reports of police action and interference at meetings, and the arrest of *Peace News* sellers.[12] Government and public alike increasingly feared Fifth Column activity, and the infiltration of groups such as the PPU by Nazi and Communist sympathizers.[13] For Vera Brittain, these doubts and suspicions cut deeply, because loyalty and trustworthiness were central to her concept of self.

The maelstrom of events around Brittain led her to a deepened awareness of the issue of trustworthiness. It now became for her a matter of living up to the demands of her own moral imperatives, which she was coming to define in terms of Christian pacifism. In November 1941, Brittain shared a crucial moment in the evolution of her thought with the readers of her fortnightly *Letter to Peace-Lovers*:[14]

> Last autumn [1940], bombed, buffeted and humiliated, I spent many hours in ruminating bitterly on the injustice of a situation in which I had, as it seemed, been cruelly deceived into parting with my beloved children before being told that, as a pacifist, I should no longer be allowed to undertake the work which the previous winter had been represented as my duty. One October day, staying with friends between Reading and Oxford, I walked along a Berkshire lane angrily meditating how easily, armed with my pen and the articulateness which is the endowment of every writer, I could one day 'get my own back' on the officials who had made such wreckage of my personal life. Suddenly, in an empty valley covered with fallen leaves, something seemed to check the direction of my thoughts. Within my mind, an inconvenient second self addressed me firmly.
>
> 'Don't you realize that this is a spiritual experience? For the past few years you have had far more honour and appreciation than you deserve. Now you know what it is to be humiliated; and this gives you a new kinship with those to whom you have hitherto felt superior – prisoners, refugees, the unemployed, the down-and-outs, and all the despised and rejected of men.' And I remembered writing four years earlier in a novel: 'I suppose, if we took a long enough view, we should feel that any sorrow bears its own compensation which enlarges the scope of human mercy. Some of us, perhaps, can never reach our honourable estate – the state of maturity, of true understanding – until we have wrested strength and dignity out of humiliation and dishonour.'
>
> One does not always get the chance to try out, in one's own life, a conclusion written with full conviction about a character in a story, and the opportunity must certainly be regarded as a privilege. I don't say that it looks like a privilege always, or even often. My thwarted love for my lost children gives me, again and again, a deep sense of injury, in which I still contemplate picturesque literary revenges on leading officials in the Ministry of Home Security. I am, unfortunately, a very imperfect pacifist and an even more imperfect Christian; and so far I have learnt only enough to recognize that this mood is wrong without always being able to avoid it . . . .[15]

Brittain's reflection on her emotions resulted in a cathartic spiritual experience, that both strengthened and deepened her pacifism. In early February 1942, Brittain wrote '[pacifism] is, in fact, the deepest religious conviction I possess. It is always, of course, easy to rationalize one's self . . . into abandoning an unpopular position, but if I did it should be, in effect, the same position as Peter before the Crucifixion, and nobody knows that better than myself.'[16] The foundation of

Brittain's spirituality, in her own words, 'comes down at rock bottom to two very familiar phrases from the Gospel – "Do unto others as ye would be done by" and "Father, forgive them, for they know not what they do" . . .'.[17]

Brittain believed that she was paying a high price for holding certain minority opinions; nevertheless, she sought to keep a balanced perspective and continually struggled to rise above self-pity. Her determination to encourage pacifists and all others that she could reach to think about peace was the spirit with which she wished to infuse *Humiliation with Honour*. In December 1940, Brittain began a correspondence with the publisher Andrew Dakers, exploring with him the possibility of writing a book about pacifism called 'We, the Minority'.[18] In it, she hoped to correct some of the popular misunderstandings about pacifists and to discuss the 'totalitarian treatment' that minorities often received in wartime.[19] By November 1941, circumstances allowed Brittain to write to Dakers:

> I now have in mind a little book which I would like to call HUMILIATION WITH HONOUR. Its object, using certain experiences of my own in a brief introductory chapter as a peg to hang the book on, would be to try and find some kind of spiritual meaning, and some source of strength, in the experiences which so many people throughout Europe and Asia are now undergoing.[20]

Her understanding of her life experience was that war had not worked against German militarism in 1914–18:

> Today we who lost our friends and lovers between 1914 and 1919 are faced with the bitter fact that all the suffering and service of those nightmare years failed completely in their purpose. Far from smashing German militarism and making the world safe for democracy, their long-range consequence has been to smash German democracy and make the world safe for militarism.[21]

This powerful message, which was also that of *Testament of Youth*, explains her perspective on the inter-war period, her perception of the present conflict and her view on its aftermath. Her analysis of the major ills of the inter-war period was a classical liberal one, rooted in what she understood to be the failures of the Versailles peace and the evils of imperialism and nationalism.[22]

Brittain was convinced that the peoples of the world needed to try something radically different. She believed that the root causes of Nazism would have to be addressed, if Hitler and all that he represented were to be defeated. She was sure that a generous peace would gain more for the world than a military victory, since the latter, she thought, would only push evil underground until it erupted again.[23] This was her analysis of the relationship between the Allied military victory of 1918 and the events of 1939.[24] A lasting peace could only be one that saw the triumph of spiritual forces, established on foundations of truly Christian values: generosity, toleration, magnanimity and understanding.[25] This is why Brittain feared the consequences of aspects of Allied strategy, notably the blockade of occupied Europe and the area bombing campaign. She believed that they would only serve to strengthen Hitler's hold 'and unite the despairing German people behind him'. Moreover, the prospect of military occupation and the imposition of another *diktat* would only further discourage an internal revolt.[26] Brittain acknowledged that it was too late for pacifism to effect the defeat of Hitler, but she insisted that it could help men and women 'to keep their heads', and possibly avoid another or even worse era in the future.[27] To do this meant to plan for peace and to ensure the foundations for it were there. Brittain's reflections in *Humiliation with Honour* were offered in this spirit.[28]

Humiliation with Honour contained her ideas about Christian pacifism, but they were ideas to which Brittain was already giving concrete expression and witness. In her view, respect for humanitarian and Christian values, and for international law would preserve a common basis of appeal, not only for all the victims of power to whom she dedicated *Humiliation with Honour*, but for all of mankind. These were the moral imperatives that propelled Brittain into her work for famine relief and her questioning of the area bombing campaign. Early in 1942, the campaign for food relief for occupied Europe became Brittain's primary focus. Her pamphlet on the subject, entitled *One of These Little Ones*, was published by Dakers in February 1943, and by July had sold 32,862 copies.[29] Brittain's work for food relief was unstinting. A scrapbook of cuttings on her lectures and public addresses covering a five-year period, from 1941 to 1946, gives compelling witness to Brittain's tireless work for the PPU's Food Relief Campaign (FRC), especially after she became its Chair in March 1943.[30]

Brittain's heavy schedule of commitments to the FRC no doubt occasioned her decided reluctance to accept an invitation, in August

1943, to write a pamphlet on area bombing for the Bombing Restriction Committee (BRC).[31] The BRC began life as the Committee for the Abolition of Night Bombing. When this aim was recognized to be unattainable, it changed into a Committee 'to urge the Government to stop violating their declared policy of bombing *only military objectives*, and particularly to cease causing the deaths of many thousands of civilians in their homes'.[32] Although the BRC's leadership was predominantly a pacifist one, the purpose of the founders was to coordinate a protest against area bombing together with non-pacifists who questioned the bombing policy of the Royal Air Force whether on legal, strategic or moral grounds. Brittain was one of the founding members of the BRC and a logical choice to undertake the pamphlet. The timing of the BRC's approach to Brittain was likely to have been prompted by the raids on Hamburg that had taken place in the summer of 1943, and which occupied many column inches in the British newspapers.[33]

Brittain's reluctance carried on into September, because of her very considerable ongoing commitments. She rather hoped that she could strike a compromise with the Chair of the BRC by encouraging him to find someone who would be able to work the material in the welter of newspaper cuttings he had sent to her 'into some kind of rough draft which I could re-write in popular form [since]. . . . [i]t is not the actual writing but the work of going through a large number of cuttings and making notes on selected items, that actually takes up the time'.[34] The Chair urged Brittain to go through the cuttings herself, many of which he had marked up, so that she might have a 'first-hand rather than a second-hand approach to the subject'.[35] In spite of all her other activities, Brittain finished the manuscript of *Seed of Chaos* in less than two months. She encompassed all of the BRC's main points of issue, drawing illustrative material almost exclusively from British newspaper sources.

Understandably, given that Andrew Dakers had already published both *Humiliation with Honour* and *One of These Little Ones*, Brittain was confident that he would publish the new work. To her surprise Dakers declined, citing an acute shortage of paper. Persisting, Brittain elicited a rather more revealing answer from Dakers, who was concerned that Brittain's book might be suppressed:

> There is very little doubt but that my list is being watched after the publication of your book [*Humiliation with Honour*] and pamphlet

> [*One of These Little Ones*] and Roy Walker's *Famine over Europe*, and this booklet is a protest, if not an attack, very powerfully stated and documented against a phase of military policy which is generally regarded as almost the most vital to success in the West.[36]

Dakers was sure that the Paper Control was in liaison with the Home Office and that if Brittain's book was suppressed, he would have his paper licence withdrawn. It is clear that his need to refuse Brittain troubled Dakers. On 10th January he wrote an unsolicited letter to her, which continued his earlier one:

> If my paper supply were cut off now, the firm would be smashed, and it is doubtful whether I would be in a position to revive it when controls disappear . . . I want you to know that I have not lost my nerve. . . . [I]f my firm has to die for its policy, I don't feel that it should yet. . . . I just want you to understand.[37]

In late January 1944, New Vision Press agreed to publish the book and Brock Printers, which produced Brittain's *Letter to Peace-Lovers*, agreed to print it.[38] Publication was set for 19th April 1944. The previous month, however, the first, shorter version of the typescript appeared in the United States under the title 'Massacre by Bombing' in *Fellowship*, the organ of the American Fellowship of Reconciliation (FoR).[39] Brittain had given a copy of the manuscript to a friend Felix Greene. Greene had submitted the article to the British censors, who had vetted it and forwarded it to him in the United States. At Brittain's suggestion, Greene had given it to Dr Nevin Sayre at the FoR, who not only prepared the piece for publication, but also arranged for a preface that was signed by 28 leading American writers and clergy in support of the arguments made in the article.[40]

Brittain was astonished that the censorship department had forwarded the manuscript to Greene, commenting that it 'says a tremendous amount for our British censors . . . and for the democratic right of British and American citizens to express unpopular opinions'.[41] She was elated at the attention her piece received in the United States. To an American correspondent she wrote: 'nothing has thrilled us [the BRC] so much as the size and seriousness of the discussion. . . . The success of the Fellowship supplement in arousing attention went, of course, beyond our wildest dreams.'[42] Her pleasure, however, was quickly tempered by the distortions of her work that soon appeared in many reviews and commentaries.[43] She was struck by the inability of most of her critics to grasp the fact that the protest was not against bombing, but against a type of bombing, namely area or saturation bombing.[44] Pre-eminent among those who failed to make this distinction was William L. Shirer, who wrote a blistering review in the *New York Herald Tribune*:

> With all due respect to Miss Brittain and her twenty-eight pacifist followers in this country, one is bound to report that many of the 'facts' in this strange pamphlet turn out to be reproductions of Nazi propaganda. In fact, Dr Goebbels, with whose writings and tricks and lies I have certain familiarity, would hardly have written it differently. . . . Miss Brittain uses Nazi propaganda to prove how frightful our bombing is.[45]

This attack was one that Brittain could not ignore. In July 1944, she protested:

> In the book *Seed of Chaos* from which my pamphlet ['Massacre by Bombing'] was selected, I made exactly ten references to direct German statements, out of literally hundreds to British officials, airmen, statesmen, newspapers, etc. . . . But minorities in war-time are accustomed to be misrepresented, and I can honestly assure you that I didn't mind – not even though you have cut down by half the sales of my more usual novels and biographies, as no doubt you have![46]

One of the few American commentaries that did call for a closer reading of Brittain's work appeared in the *Catholic World*. James Gilles, the Editor, believed that Brittain's work 'deserved either endorsement or patient refutation'. He had been travelling and had followed the American papers 'in half a dozen states and twice as many cities', so he was able to report that: 'I must have read a couple hundred attacks on Miss Brittain and her theme. They ranged all the way from expostulation to diatribe, but I'm blessed if I found one which correctly reported the precise position she had taken.'[47]

President Roosevelt's widely published rebuke to the 28 signatories and to Vera Brittain began by stating that the President was 'sorry that he cannot agree either with the "facts" or with the conclusions of the article'. He finished by asserting:

> You cannot talk conciliation. You can be for the Germans and the Japanese and look forward to new 'Dark Ages' attended by

> world-wide death and destruction – or a continuance of the philosophy of peace and the maintenance of civilization. You cannot effect a compromise between these two views.[48]

Brittain's response was, 'I must take Christ as my guide before even the President.'[49] To the Quaker founder of the BRC, T. Corder Catchpool, she wrote that the President's 'comments seem to me largely an attempt to side-track us, and make me and the 28 say things we never did say. . . . I too now feel (especially since Roosevelt intervened) that Shirer was put up to discredit our effort.'[50]

The publication of *Seed of Chaos* in Great Britain in April received much less attention. Corder Catchpool had hoped to be able to capitalize on the publicity in the United States, but he reported to Brittain with disappointment that George Bell, the Bishop of Chichester, who in February 1944 had spoken against area bombing in the House of Lords, and who had seen some of the American newspaper cuttings, 'refused to be one of the speakers at [a BRC] meeting, and even refused to send a message to be read' at the gathering.[51] A month later Catchpool had to report that his efforts to encourage the Bishop to support a collection of signatures, along the lines of the Fellowship's 28, had failed. 'I am inclined to think that the omission of his name would appear so significant to other potential signatories and to the general public that we may have to abandon the idea.'[52]

Other bad news was also recorded. The *British Weekly* had reprinted Shirer's attack on 'Massacre by Bombing' on 4th May, but it steadfastly refused to allow Brittain an opportunity to reply. Accordingly, H. S. Jevons, Chair of the BRC, and Catchpool wrote to the Editor of the paper and stated that they had supplied most of the material quoted by Brittain and that 'on behalf of the Committee [they took] primary responsibility for its use'. They requested that the author be allowed to answer the 'serious and unsubstantiated charge, so damaging to . . . [her] reputation as a writer' so that the readers of the *British Weekly* could 'form their own judgement'.[53] Their appeal fell on deaf ears. Savage criticism was also forthcoming from George Orwell, who did acknowledge the distinctions Brittain endeavoured to make in her piece, only to dismiss them abruptly: 'She is willing and anxious to win the war, apparently. She merely wishes us to stick to "legitimate" methods of war and abandon civilian bombing, which she fears will blacken our reputation in the eyes of posterity.' Orwell, who wrote that 'pacifism is a tenable position', believed, nevertheless, that 'all talk of "limiting" or "humanizing" war, is sheer humbug'.[54] Nor did Brittain's piece find support in either *Peace News* or the Quaker journal *Friend*. Both gave the campaign short shrift in their columns because of the absolutist pacifist stance of their respective editors on the bombing issue.[55] It was a range of reactions that proved prophetic.

Brittain herself reflected at the end of the war that her writing represented 'a small part of democracy's universal resistance-movement against the all-powerful state, and a characteristic reaction of the individual and independent mind against the totalitarianism of our time'.[56] But beyond that, *Seed of Chaos*, which had put Brittain at the centre of a fierce exchange of views in the United States, was taking its place within a larger debate on the issues of bombing, a debate which has continued ever since, its tides of judgement ebbing and flowing over time.

Acknowledgement

My thanks to Alan Bishop, Kathleen Garay, Robert Goheen, Frances Montgomery and Irene Sanna – Y.A.B.

Notes

1 Vera Brittain, *Testament of Youth. An Autobiographical Study of the Years 1900–1925* (London, 1933). Brittain's life story is continued in her biography of Winifred Holtby, *Testament of Friendship* (London, 1940), and in *England's Hour* (London, 1941) and *Testament of Experience* (London, 1957). There are two biographical studies of Brittain: the authorized biography, Paul Berry and Mark Bostridge, *Vera Brittain: A Life* (London, 1995) and Deborah Gorham, *Vera Brittain: A Feminist Life* (Oxford, 1996), which covers her life up to 1939.

2 Canon Dick Sheppard to Vera Brittain (hereafter VB), 27 January, 1937. Hereafter all the correspondence to and from VB comes from the Vera Brittain Collection (VBC), William Ready Division of Archives and Research Collections, McMaster University, Hamilton, Ontario. In the same letter Sheppard asked VB also to become a Sponsor of the Union. Amongst pacifist groups the Government recognized the pre-eminence of the Peace Pledge Union, not only in numerical terms, but also as the pacifist body most actively and

visibly opposed to the war. CAB 73/3 CDC (40) 8, 'Home Front Propaganda', Memo by the Minister of Information, 3/3/40. VB became a sponsor of the PPU in February 1937 and served as a member of its Executive Council.

3 'Peace Leaflets Distributed: Protests by Recipients', *The Times*, 7 October, 1939, p. 3 col. c; Vera Brittain, 'The Twilight of Truth. Vera Brittain discusses the misrepresentation of the Peace Pledge Union', *Peace News*, 27 October, 1939, p. 4; 'Communazi 5th Column In Britain', *Evening Standard*, 15 April, 1940, in file marked 'annotated holograph manuscript', Humiliation with Honour VBC/A11. Also VB to Margaret Storm Jameson [MSJ], 15 October, 1939; VB to MSJ, 4 August, 1940. For a full discussion of the wartime Peace Pledge Union, see Yvonne Aleksandra Bennett, 'Testament of a Minority in Wartime: The Peace Pledge Union and Vera Brittain, 1939–1945', PhD dissertation, McMaster University, 1984.

4 VB to Ruth Colby, 1 February, 1942. Ruth Colby was an American friend of Brittain.

5 MSJ to VB, 19 April, 1940; 29 January, 1941; 5 May, 1941; 6 September, 1941.

6 Vera Brittain, 'What Can We Do In War Time?', *Forward*, 9 September, 1939, VBC/G543. Also quoted in Alan Bishop and Y. Aleksandra Bennett (eds), *Wartime Chronicle, Vera Brittain's Diary, 1939–1945* (London, 1989), pp. 18–21.

7 Brittain, 'What Can We Do In War Time?', *Forward*, 9 September, 1939.

8 Vera Brittain, *Humiliation with Honour* (London, 1942), also published by Fellowship Publications (New York, 1943). *Seed of Chaos* (London, 1944). A shortened version, under the title 'Massacre by Bombing', appeared in *Fellowship*, vol. X, no. 3, March, 1944, pp. 49–64. This was also available as a pamphlet (March, 1944).

9 VB to MSJ, 18 January, 1939.

10 VB to James Hudson, 12 December, 1938.

11 Lectures and speeches (England). September, 1933–December, 1940. VBC/E20.

12 *Peace News*, 8 March, 1940, p. 8; 15 March, 1940, p. 8. *Peace News* was the weekly newspaper of the Peace Pledge Union.

13 VB was also very upset at the Fifth Column drive of the London *Evening Standard* newspaper: VB to MSJ, 18 May, 1940. Also VB to George Catlin, 19 October, 1940. *Peace News*, 1 March, 1940, p. 1, 12 and also 15 March, 1940, p. 2. House of Commons *Debates*, vol. 357, col. 1505–6, 22 February, 1940. Neil Stammers, *Civil Liberties in Britain During the 2nd World War* (Beckenham, 1983).

14 VB began issuing her fortnightly newsletter *Letter to Peace-Lovers* (LPL) on 4 October, 1939.Within a year, it had been singled out in a

government memorandum as one of the most important publications of its sort (PRO CAB 75/7 HPC (40) 103, Anti-War Publications, 4 May, 1940). It was her intention to provide interested and like-minded individuals with an opportunity for fellowship and a forum in which problems and ideas relating to conflict and peacemaking might be discussed. *Humiliation with Honour* and *Seed of Chaos* contained many of the ideas and themes that VB first raised and worked out in LPL as well as in private correspondence. Winifred and Alan Eden-Green edited a selection of the *Letter to Peace-Lovers* published under the title *Testament of a Peace Lover* (London, 1988).

15 LPL no. 70, 20 November, 1941, VBC/A12. On 26 June, 1940, VB's children, John (aged 12) and Shirley (aged 9), had been evacuated to the United States, as France collapsed and the threat of invasion loomed large. VB and her husband George Catlin had taken the painful decision, believing that it would be possible for the children to be visited. The departure was a decision that she almost immediately regretted. See Berry and Bostridge, *Vera Brittain*, pp. 386–424.

16 VB to William Kean Seymour [WKS], 8 February, 1942 [Matthew 26.70–4]. Seymour was a writer and Brittain's bank manager.

17 VB to WKS, 18 February, 1942 [Matthew 7.12; Luke 23.34].

18 Andrew Dakers [AD] to VB, 10 December, 1940. In 1939, Dakers had established his own publishing house with the express purpose of taking up the 'causes of peace, goodwill, social security, and sane domestic and international relations'. AD to VB, 10 December, 1940, and 16 December, 1940. He was not a pacifist.

19 VB to AD, 12 December, 1940.

20 VB to AD, 19 November, 1941. To this end VB hoped that it would be possible to produce an inexpensive edition in order that it might reach a wider audience. In letters to Agatha Harrison (19 November, 1941) and Ruth Colby (1 February, 1942), Brittain wrote that the book was a consequence of the year's previous experiences. At the time of writing, she believed that the Home Security Department, unable to secure her imprisonment, sought to undermine her personal and professional life in so far as it could.

21 Brittain, 'What Can We Do In War Time?', *Forward*, 9 September, 1939.

22 See J.M. Keynes, *The Economic Consequences of the Peace* (London, 1919) and Harold Nicolson, *Peacemaking* (London, 1933) for two classic interpretations of the peace very familiar to Brittain. The widespread disappointment and disillusionment with Versailles in liberal and socialist circles in Britain, alongside protest in Germany, contributed to the successful undermining of Article 231 of the Versailles Treaty (the war guilt clause) by successive Weimar governments. This in turn helped to create the intellectual and historical climate for

appeasement. See Holger Herwig, 'Clio Deceived: Patriotic Self-Censorship in Germany After the Great War' in Keith Wilson (ed.), *Forging the Collective Memory: Government and International Historians Through Two World Wars* (Oxford, 1996); C.A. Cline, 'British Historians and the Treaty of Versailles', *Albion*, vol. 20, no. 1, 1988, pp. 43–58.

23 VB to WKS, 27 February, 1942. This is also why, in the first part of the war, VB campaigned for a negotiated peace. Bennett, 'Testament of a Minority in Wartime', pp. 307–24.

24 VB was shocked by those pacifists who offered excuses for Nazis. For her there were no excuses, rather explanations. See VB to Monica Whately, 8 December, 1941.

25 VB to George Brett [GB], 20 March, 1942. Brett was President of Macmillan in New York.

26 VB to GB, 20 March, 1942. Again, Brittain's analysis is informed by her understanding of the consequences of the 1914–18 war.

27 VB to MSJ, 10 May, 1941.

28 In a letter accompanying the finished manuscript, VB commented to Dakers that war-time book reviews were, in her experience, an easy and proven way of allowing government-inspired reviewers to sabotage the works of dissenting authors. In Sweden and the United States, *England's Hour* (1941) had received reviews rivalled only by those of *Testament of Youth*, 'but here [she reported] it was ruthlessly slaughtered by several critics of whom I could identify at least a few as members or ex-members of the Ministry of Information', VB to AD, 4 August, 1942. *Humiliation with Honour* was more noticed than reviewed, which makes the sales it achieved all the more remarkable. Dakers jubilantly reported to VB that he had 'more than 3,000 orders in hand before the book has been subscribed in London at all', AD to VB, 17 October, 1942. The book had a drawing by Arthur Wragg on the jacket that VB had commissioned and paid for herself. In her instructions to Wragg she wrote: 'What I want is something which, like the book itself, will emphasize not so much the horror and suffering in the world but the spiritual grace by means of which people can not only be victorious in their own disasters, but help others in a similar situation.' VB to Arthur Wragg, 23 January, 1942.

29 AD to VB, 12 July, 1943. Dakers reported to VB that William Temple, the Archbishop of Canterbury, had in his possession material from *One of These Little Ones* for his address to both Houses of Parliament, but that he 'doesn't want it publicly associated with his address . . . on Feb. 17th [1943] because of its mild and vague criticism of the Gov.!' *Wartime Chronicle*, p. 212.

30 Day-by-day book 1940–51, VBC/E23; Day-by-day book January, 1941–December, 1946, VBC/E24; Yvonne Aleksandra Bennett, 'A

Question of Respectability and Tactics: Vera Brittain and Food Relief for Occupied Europe, 1941–1944' in Harvey L. Dyck (ed.), *The Pacifist Impulse in Historical Perspective* (Toronto, 1996), pp. 384–96.

31 H.S. Jevons [HSJ] to VB, 25 August, 1943.

32 Bombing File – Bombing Restriction Committee, *Stop Bombing Civilians!* (London, 1943), VBC/E34.

33 Martin Middlebrook, *The Battle of Hamburg: Allied Bomber Forces Against a German City in 1943* (New York, 1980); Noble Frankland and Charles Webster, *The Strategic Air Offensive Against Germany 1939–1945*, 4 vols (London, 1961).

34 VB to HSJ, 13 September, 1943. VB had a number of speaking engagements and was also expecting her daughter to return from the United States. John had returned home on 18 July, 1943. About Shirley's return, on 17 October, 1943 VB wrote: 'I felt as though, after years of unbearable climbing, I had reached the summit of my Everest.' *Testament of Experience*, p. 322.

35 HSJ to VB, 16 September, 1943.

36 Dakers continued: 'I would willingly take any unreasonable risks to publish the booklet, but I cannot include others, such as the authors whose books I have contracted to publish, to share a risk towards which they may not even be sympathetic. A court action and a fine is one thing, but the debarring one from continuing one's business is very much another.' AD to VB, 29 December, 1943; Roy Walker, *Famine Over Europe* (London, 1941).

37 AD to VB, 10 January, 1944. VB wrote a very understanding reply to Dakers on 12 January, 1944.

38 VB to Alex Comfort, 25 January, 1944. The delay in finding a publisher and a printer 'who dared to undertake it' resulted in the book's expansion, since VB wanted to keep the piece up to date: VB to Stephen Hobhouse, 8 February, 1944 and VB to GB, April 16, 1944. The book's title and the illustration for the jacket cover were subjects of some discussion. The picture on the original jacket was drawn by Pleasance Catchpool and was probably based on a photograph of a raid on Nuremberg published in the *Daily Telegraph*. See VB to P. Catchpool, 15 February, 1944, and HSJ to VB, 16 and 21 September, 1943; VB to GB, 16 April, 1944.

39 Vera Brittain, 'Massacre By Bombing', *Fellowship*, vol. X, no. 3, March, 1944, pp. 49–64.

40 Copy of letter from Felix Greene to T. Corder Catchpool, 10 March, 1944, VBC. Also VB to GB, 16 April, 1944, and VB to Dorothy Detzer [DD], 18 April, 1944.

41 VB to GB, 16 April, 1944.

42 VB to DD, 18 April, 1944. To Detzer she commented: 'It was almost by accident that the manuscript got to the USA at all.' The *New York*

Times alone devoted considerable space to the debate arising from the publication of 'Massacre by Bombing': 'Obliteration Raids on German Cities Protested in U.S.', 6 March, 1944, p. 1, 11; 'Massacre by Bombing', 8 March, 1944, p. 18; 'Letters to the Times: the Bombing of Germany', 8 March, 1944, p. 18; Anne O'Hare McCormick, 'Abroad: The Aerial Invasion of the Continent', 8 March, 1944, p. 18; 'Letters to the Times: More on Bombing', 9 March, 1944, p. 16.

43 Day-by-day book January 1941–December 1946, VBC/E24; Bombing File VBC/E34 and *Seed of Chaos* VBC/A13. See also Conrad C. Crane, *Bombs, Cities and Civilians: American Airpower Strategy in World War II* (Lawrence, 1993), pp. 28–35; Stephen A. Garrett, *Ethics and Airpower in Word War II: The British Bombing of German Cities* (New York, 1993), pp. 120–7. There are few other books in the voluminous literature on bombing that recognize VB's contribution to the area bombing debate.

44 VB to Mrs Glen, 21 April, 1944.

45 William L. Shirer, 'Propaganda Front. Rebuttal to Protest Against Bombing', *New York Herald Tribune*, 12 March, 1944, VBC/E24; VBC/ E34.

46 VB to William Shirer, 19 July, 1944.

47 James M. Gillis, 'Editorial Comment', *Catholic World*, Vol. CLIX, May, 1944, pp. 97–104, VBC/ E34.

48 'Mass Bombing Foes Rebuked By Roosevelt', *New York Herald Tribune*, 26 April, 1944. The subtitle of the article boldly declared 'Their "Facts" Are Wrong'. G.E. Neyroud, 'F.D.R. Defends Mass Raids', *News Chronicle*, 27 April, 1944, VBC/E24.

49 VB undated draft, VBC/E24.

50 VB to T. Corder Catchpool (CC), [n.d.] April, 1944.

51 CC to VB, 23 April, 1944. Bell's address was published in pamphlet form by the BRC in 1944 under the title *The Bishop of Chichester on Obliteration Bombing*, VBC/E34. In November 1943, however, he had refused to sign an open letter to the Government opposing mass bombing. M.J. Douglas to Vera Brittain, 9 February, 1944.

52 CC to VB, 20 May, 1944.

53 CC and HSJ to John Hutton, 18 May, 1944.

54 George Orwell, 'As I Please', *Tribune*, 19 May, 1944, VBC/E24. See Alan Bishop, 'Vera Brittain, George Orwell, Mass Bombing and the English Language' (paper presented to McMaster English Association, McMaster University, February 1984). Also Brittain, *Testament of Experience*, p. 331 and CC to VB, 21 June, 1944. The President's rebuke was reported in Britain. See G.E. Neyroud, 'F.D.R. Defends Mass Raids', *News Chronicle*, April 27, 1944, VBC/E24.

55 Corder Catchpool was deeply distressed by the review in the *Friend*: CC to VB, 21 June, 1944.

56 Foreword to *One Voice, A Collection of Letters* by Vera Brittain, Letters to Peace-Lovers, VBC/A12. It is worth noting that Brittain found herself, together with Churchill, on a Gestapo list (*Sonderfahndungsliste GB*) of 'some 2,000 persons whose arrest was to be "automatic"' following the successful occupation of Britain. Her husband, George Catlin, was also on the list. Vera Brittain, *Testament of Experience*, p. 388.

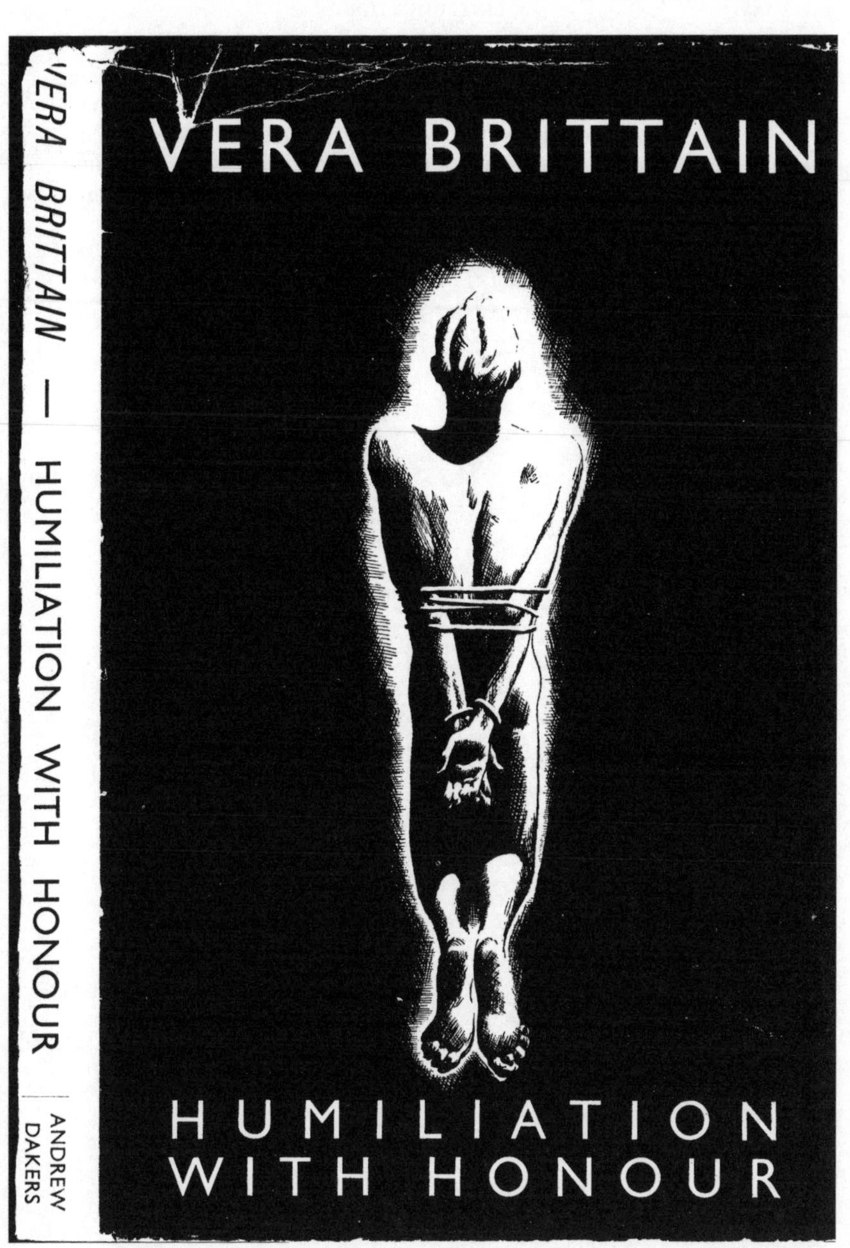

Original cover of *Humiliation with Honour* by Arthur Wragg, first published in 1942 by Andrew Dakers Limited, London.

Humiliation with Honour

Vera Brittain

To the victims of power

It is only the experience of historical failure itself that has proved fruitful, in the sense that the consciousness of humanity has thereby been increased.

Nicolas Berdyaev, The Meaning of History

Be not ashamed, my brothers, to stand before the proud and the powerful
With your white robe of simpleness.
Let your crown be of humility, your freedom the freedom of the soul;
Build God's throne daily upon the ample bareness of your poverty,
And know that which is huge is not great and pride is not everlasting.

Rabindranath Tagore, The Sunset of the Century
(from Nationalism*)*

Prologue

My Dear Son, In one of R.'s recent letters, she told me that you and S. had been talking to her about the War and my attitude towards it. S., she said, had championed, with the passionate loyalty so characteristic of her, the ideas of the minority to which I belong. But you had questioned and criticized.

Let me say at once that I find it both right and natural that S. should applaud and you should criticize. Being nearly three years your junior, she still is younger than you were when you both went to America. At that time, you had not yourself reached the age of philosophic interrogation. But you have grown up so quickly in the past two years that your questions are now inevitable.

I need hardly say that I do not expect you, at fifteen, to endorse any beliefs that I hold just because they are mine. No one has much value as a member of a minority unless he has once thought with the majority, and shared the position of those who now call him a 'fanatic' or a 'crank'. What matters most at your age is that you should think for yourself.

But it is also important that you should understand the thinking of others. Your criticism will not be authentic if it is directed against fake images and imaginary assumptions, and you will find that most wartime newspapers and magazines will tend to create those images and to make those assumptions. You must have a clear idea of what some of us really think about the world we are now living in, before you condemn us for our point of view and the penalties which it brings. Above all, I want you to realize that even humiliation is not dishonourable when it is voluntarily accepted and endured without bitterness for the sake of convictions which those who hold them believe to be true.

Pacifism is nothing other than a belief in the ultimate transcendence of love over power. This belief comes from an inward assurance. It is untouched by logic and beyond argument – though there are many arguments both for and against it. And each person's assurance is individual; his inspiration cannot arise from another's reasons, nor can its authority be quenched by another's scepticism.

In my next few letters I shall try to explain what I mean by these assertions. My letters will not be designed to solicit your agreement. Their purpose is only to discuss a minority problem made acute by war, and to illustrate a point of view which is shared by many people in addition to

Your Mother

I

The Wheels of Juggernaut

I do not think I have ever told you just when and how those beliefs which you question began.

Their origin lies far back in the last War, which ended nine years before you were born. For two months of that War, during 1917, I was nursing wounded German prisoners in a camp in France.

Before that, I had felt much the same about Germans as most people feel today. Though I couldn't, even then, swallow quite all the propaganda that I read, I took for granted, like other young people of my generation, that the Government was always right, and that it was our duty to kill Germans because they said so.

Those two months started me thinking for myself in much the same way as you are thinking now. Before they were over, they had taught me that the qualities common to all human beings, of whatever race or country, far exceed the national and political differences which sometimes divide them.

That discovery was one of those profound spiritual experiences which make all controversial argument seem trivial and irrelevant. Even after 25 years I have still to learn its full implications, though it made me resolve to devote my life to examining the causes of war and doing what I could to prevent another. This endeavour has failed for the time being; but I should be false to myself, and you, and everything which matters to us both, if I were to deny today what I learned from that spiritual experience.

Some of us who remember the First World War recall their oppressive consciousness of a vast machine gradually taking control of the men who made it. The politicians of the Great Powers had set the wheels of Juggernaut in motion, but almost immediately the chariot seemed to take its own direction, and to roll over humanity at its self-chosen pace.

Today, in a second and still more destructive War, the same thing is happening all over again and on an even larger scale. Because they proved incapable of learning the lessons of history, some of the blind leaders who thought that they could control the machine are already numbered amongst its victims. The speed of the crushing wheels which they started is carrying us to the end of that epoch which began four centuries ago with the Renaissance and the Reformation.

I expect you often see the American news magazine *Time*. At the beginning of 1942[1] there was a sentence in it which has stayed in my mind: 'In 1941, over the world's measureless acres of misery, the War lay like a burden too great to be carried, too great to be thrown off.' That sentence referred to the period before Pearl Harbour. In 1942 the burden grew heavier, and the acres of misery multiplied.

If you turn your eyes to the world as a whole from the land of rich resources where you are now a guest, you will see a picture of suffering probably unparalleled in history. The worst peace conceivable would not have produced a tenth of it.[2] In spite of its darkness this picture contains its elements of hope, but I shall write of these in later letters. Today I want you to realize the measure of unnecessary pain which has come to mankind through a war that could have been prevented on a score of occasions if those in authority had really intended to do good instead of evil. This deliberate choice of the worse way rather than the better was typical of the nations which won the last War, long before it began again to be typical of Germany, or of Japan.

The peoples who are suffering today still have more reasons for liking than for detesting one another, just as the British Tommies and the German prisoners found that they had in 1917. But their rulers, each of whom is playing his own game of power-politics and blaming the ensuing catastrophes upon the others, do their best to prevent them from discovering the true extent of their affinity. By emphasizing the latent antagonisms which can divide communities, and attributing to all their opponents whole categories of qualities which only a few of them possess, they persuade ordinary men and women to hate, despise, and kill others of their kind with whom they have no quarrel.

Never, within so short a period, have great populations all over the world been tortured in such a variety of ways both mental and physical. We know this to be happening, though one of the unique features of this War is that we seldom receive any trustworthy details. The shapes which emerge from the kaleidoscope of pain are dim and confused. Like mist concealing the outlines of mountain summits, the fog of propaganda blankets the sharp edges of truth.

Here in Europe, we have seen the political achievements of millennia fall into dissolution. In Asia the British and Dutch Empires, built through centuries by the toil of millions, disintegrated in three months. Every morning we in England and you in America read in our newspapers of events comparable to the clash of planets, of historic changes greater than those which caused the fall of the Roman Empire. But have you ever noticed how little we learn of the day-by-day incidents which give human meaning to these revolutionary events? What actually lies behind the reassuring platitudes of the radio or between the lines of the Press reports, so carefully drawn up in words designed to neutralize tension and reduce earth-shaking cataclysms to the comfortable proportions of normality?

We do not know. Because the British, like the Americans, are a humane people, the barrier of censorship has to be imposed and the news sifted all the more carefully. We recoil from the thought that, by blockading the Continent, a government which we elected is conniving at the starvation of millions. So we must not be permitted to picture the poor homes and ill-equipped, overcrowded hospitals in which Greek and Belgian children are dying from famine or the diseases which it brings. Nor must we be reminded of the aged and the invalids who shuddered through the last bitter winter without food, heat, or warm clothing.

The imagination of kindly men and women is protected against these facts because, if they were aware of what is really happening behind the drawn curtains, their decency would revolt. If they could see starvation, disease, and death actually at work, they might begin to ask, on an embarrassingly large instead of a tiny scale, whether any victory was worth having which involved the continuation of such methods. They might also remember that the makers of international law, now partially repudiated by both sets of combatants, based their rules upon the doctrine that it is better to suffer disadvantage in war than to descend to conduct so barbarous that it strips the cause in whose name it is committed of even the bare pretence of morality.

When I recollect walking through the snowy streets of Calais on the first stage of a roundabout journey to the United States in January 1940, it is strange to think that the coast of France might be as distant as mute Batavia for all that we can now learn of the daily lives of its inhabitants. You and I cannot even guess what unexpected friendships may be growing up between human creatures drawn together by their dire need of comradeship and comfort. A million examples may exist of mercy, forgiveness, and cooperation, but they pass unrecorded. Because the curtain is seldom lifted on anything but hatred and murder, we receive from Europe and Asia a dark impression of pain unrelieved by pity.

You remember the Four Horsemen of the Apocalypse, those gaunt spectres carrying tragedy through the world? We might well invent a Fifth, and call his name Silence. Today silence is the symbol of terror and bereavement, of lifelong speculation following months of suspense. A ship sails – destroyer, merchantman, trawler, or submarine. It never reaches its destination. Where is it? Silence. A bomber goes out on a night flight or a daylight sweep, and 'fails to return'. What has happened to the pilot – to my son, my husband, my brother? Silence. A small garrison holding an island fortress is beleaguered and attacked till communications cease. Why did they cease? Silence.

Perhaps we inquire what befell the civilians in other fallen cities and islands. You will remember that, for one moment, a searchlight played vividly upon a corner of Hong Kong. We averted our eyes, but ask the more urgently: what happened in Wake, in Guam, in Manila, in Penang,[3] in Singapore, in Java, in Tobruk, in Rostov? How did Mandalay appear when the Japanese had finished with it? What has occurred in Burma since it vanished from the news? Silence. In a world where the means of communication are swifter and more numerous than was even conceivable to the imagination of our grandparents, Silence has become the twin brother of Giant Despair.

Such is civilization in the fifteenth year of your too eventful life. If you mean to leave it better than you find it, you and your generation will have plenty to do. It will be worth doing. But it will not be easy for the artist into which you are developing. An artist is always a strong individualist, and the tendency of the present age is to repress the individual and impose conformity.

Next week I will tell you what I mean by this, and how it is done.

Notes

1 January 5th, 1942.

2 Nothing illustrates this fact more strikingly than the fate of the Jews in Nazi-occupied Europe. The World Jewish Congress recently stated (*Evening Standard*, August 6th, 1942) that of the seven million Jews who normally live in these territories, one million have been done to death. Yet according to a provisional census taken in May, 1939, and quoted by the Royal Institute of International Affairs, the total number of Jews in the pre-war Reich was no more than 339,892. This figure included Austria and Sudetenland, but not Danzig and Memel.

3 In a letter to *The Times* of August 14th, 1942, on 'Colonial War Damage', Mr L. D. Gammans, M.P., has stated: 'There will be damage to private property as well. Thousands of houses will have been destroyed in actual combat or by bombs. Penang, for example, is little more than a shambles . . . The houses and shops are almost entirely Asiatic-owned, and it is not always realized that over 50 per cent of the rubber and 40 per cent of the tin mines are owned and operated by Asiatics.' Mr Gammans' letter was written to inquire who is going to pay for all this damage.

2

The Decline of the Individual

In my last letter I tried to give you a general picture of the age of spiritual retrogression in which we live. Do not accuse me of pessimism because the picture was dark. I am so far from being a pessimist that even two wars have not impaired my faith in humanity and its future, though you are now more likely than myself to see the civilized society of which I have dreamed.

But wishful thinking and lack of realism are the bane of our day. It is natural that the inheritors of a terrible epoch should indulge in mental escapism, and should be encouraged in this by those who have led them to catastrophe. There is no antidote to discontent so effective as an easy sentimental optimism. But this has nothing in common with faith and courage, which spring from a knowledge and acceptance of the truth. You remember the words of Jesus to his disciples: 'Ye shall know the truth, and the truth shall make you free.' In *De Imitatione Christi*, Thomas à Kempis carried this promise to its next logical stage: 'If the truth shall have made thee free, thou shalt not care for the vain words of men.'

The vainest words are those which blunt the edge of our courage by depicting our circumstances as better than they are. We can save ourselves only if we honestly recognize the extent to which retrogression has begun, combat its manifestations with all our strength, and use whatever creative force we possess to reverse certain existing social tendencies.

If we fail in this, the human race may not recapture for centuries even the partial allegiance to spiritual values which has been characteristic of the more civilized in both East and West. I believe that our success depends largely upon our ability to reassert and recover the value of the individual as such.

When you study the history of civilization, you will find that it is the story of the individual's emergence from the control of powers hostile to his development: forces of nature, of superstition, of ignorance, of powerful tyrants and the all-powerful State. Christianity itself is based upon respect for the individual as the possessor of an immortal soul more valuable, in the eyes of God, than the material ambitions of competitive governments. Even politicians, whose vested interests are essentially anti-Christian, occasionally recognize this paramount claim.

I recall, for instance, a statement made during the last War[1] by Lord Hugh Cecil to the House of Commons, which does not often get the opportunity of hearing declarations of religious faith by its Members. He was speaking against an amendment to the Reform Bill, disenfranchising conscientious objectors, and this is what he said:

> I am most anxious that this country should maintain the proposition that there is a higher law, and that we will not listen to the doctrine that the State's interest is to be supreme . . . Belief in the State cannot help us to bear the sufferings or control the passions of the War. It is a barren faith, as well as a degrading faith. It does but encumber us and shut us out from that higher world in which we ought to live.

But today, in another War, when the need to control passions and bear suffering is even greater, Christianity itself is in retreat. And with it has declined, during the past two decades, the value of its essential unit, the individual soul. A few months ago I read a newspaper article[2] in which the writer quoted some significant words used by Joseph C. Harsch, formerly the Berlin correspondent of the *Christian Science Monitor*:

> The force which defeats Germany must also possess one essential of the German military machine. It must be backed, as is Germany's, by a Government which can exact sacrifices from the civilian mass behind the Army beyond anything America or Britain has ever faced.

Note that phrase 'civilian mass'. It implies a bulk population without individual hopes, fears, choices, affections, or any rights but the right to make more and more sacrifices of everything that renders life worthy to be lived.

Today, across the five continents and the seven seas, you find large sections of that bulk population being subjected to forms of persecution, oppression, victimization, and constraint which were used in the period between the World Wars by the Russian and German totalitarian governments, but which the unleashing of war-time passions has multiplied and extended a hundredfold. These forms of persecution range from the state control of individual travel, to the compulsory emigration of entire populations; from the exercise of autocratic powers to detain 'suspects' without trial in prisons and concentration camps, to the imposition of military and industrial conscription; from the seizure or regulation of private funds, to arbitrary restrictions on speech and writing. The one extreme passes easily into a policy of murder which views with indifference the death of the body; the other involves a total disregard of the frustration and atrophy of the mind.

Perhaps you have noticed that, since the War began, a new kind of standard vocabulary has been developed which seems to be designed to conceal the individual suffering caused both by war-time adversities and by the official use of humiliation as a weapon. I wonder if America too has begun to use this vocabulary since she came into the War? One form which it takes in our own country is the multiplication of categories which blunt the perception of personal disaster by means of a neutral-tinted, collective impersonality.

With the help of this convenient verbal system, the wounded and dying are transformed into 'casualties', while the small householders who lose the shelter and savings of a lifetime in an air raid become 'the homeless'. Those hunted individuals who flee their countries seem somehow less desperate when they are described as 'refugees'. The mothers and children who escaped under bombs from Singapore lost much of their pathos as soon as they were included in that familiar category of waifs and strays, the 'evacuees'. When a fight against overwhelming odds means a heavy sacrifice of personnel for the R.A.F. we learn to accommodate ourselves to the loss of '42 machines'. If the sacrifice is exacted from the enemy, our imaginations are helpfully soothed by the use of sport metaphors: the crashing aeroplane whose flames consume its living cargo of human flesh becomes part of a 'score' or a 'bag'.

You can see how important it is for the makers of war that men and women should guard themselves, or be guarded, against the sensitive response of their own humanity to the humanity of others. Usually they take to the process of adaptation readily enough. It is only when they have had some unforeseen and unfamiliar experience, like mine with the German prisoners in France, that they are startled out of acquiescence into realization.

In case adjustment should prove unexpectedly difficult, the standard vocabulary has a useful selection of disparaging words to assist the pupil. The best examples are usually to be found in the Sunday Press, where they are perhaps introduced with the idea of assuaging any misgivings which might have been aroused by the reading of Epistle and Gospel at morning prayer. In the pages of these newspapers, the advocates of mercy, even to our starving ex-Allies, become 'sentimentalists', while those who stop short of hatred and murder are described as 'squeamish'. As for our opponents, they never state a case or make a protest; they always 'scream' or 'bellow'. People who retain sufficient detachment to distinguish between the Nazi leaders and such Germans as the Bishop of Münster and the pastors of the Confessional Church, are dismissed as 'mushy'. The Editor of a provincial newspaper[3] once derided me as 'mealy-mouthed' because, in the hope of tempering the passions roused by an acrimonious local correspondence, I had endeavoured to make a balanced and unprovocative statement of the pacifist position. In his view, its presentation should have had more 'bite' and 'punch'!

These nouns and adjectives are emotional words, carefully chosen by journalists who are experts in verbal effects. I need hardly remind you that the prostitution of our more rational qualities is not achieved by calling them contemptuous names. What *is* prostituted is the mind of the person who confers those names or accepts them.

If you do not keep your intelligence perpetually wide awake, you will find these repeated suggestions beginning to affect your judgement in spite of yourself. The one thing that war propaganda avoids is the direct honesty of a recognizable challenge. So much of it is done by disparaging innuendoes that when you begin to feel passionate and excited, you can hardly ever recall what it was that first put your reason to sleep and roused your emotions.

I believe that the irrational state of mind created by this type of verbal propaganda is as great a danger to our nation, and its chances of winning what most people believe it to be fighting for, as Hitlerism itself. In the Nazis and the Japanese we recognize cruelty when we see it, yet that same cruelty is being created, unperceived, amongst ourselves by our constant denunciation of the very qualities – chivalry, compassion, toleration, kindness – for the absence of which we condemn our enemies. The quality now officially glorified is 'toughness' – a dangerous characteristic for a nation to cultivate. 'Toughness' may be only the younger brother of ruthlessness, but it has a habit of growing up.

By reading simply the official information which appears in the Press, you cannot help but realize that suffering is now so widespread that the disregard of it has become a policy. Even the official lists of casualties, still printed on the less conspicuous pages of British newspapers, have disappeared in many countries. Some time ago I read a dispatch sent in by the Lisbon correspondent of *The Times*,[4] describing a disaster which overtook the Spanish 'Blue Division' supporting the Axis on the Russian Front. 'Care has been taken in Spain not to publish the casualty lists,' he wrote. 'Even relatives of the dead are not informed until many weeks have passed, when a pathetic bundle of clothes arrives by post.'

The Spanish people are not alone in being deprived of the human privilege of mourning their sons. In our own democratic country, as for many years in the totalitarian states, mothers, wives, fathers, and brothers must 'take it' as unobtrusively as possible. The child who was the joy of a home may be crushed to death in an air raid; the boy whose promise had been his parents' inspiration may disappear in a futile attack on an invulnerable objective by an antiquated aeroplane. But the bereaved must not complain for fear of diminishing 'morale'.

The whole tendency of modern war propaganda is to persuade the public that 'morale' is a synonym for morality. Actually the two are far from identical. 'Morality' signifies the social interpretation of the highest ethical values; 'morale' is a false god in whose name the warring peoples of the earth risk the loss of their own humanity. Yet the very attempt, universally made, to achieve the uncomplaining endurance of the intolerable confers upon these peoples common qualities and mutual interests which cast a revealing light upon the dressed-up grievances of political leaders the moment that it is possible for the peoples to get together and discuss them frankly. If men and women remained constantly aware of the humanity in themselves and their counterparts, no war would last for an hour. One of our strongest weapons in the War against war – which mankind must win or be annihilated – is the recovery of that human awareness against which all war propaganda is directed.

Fortunately the individual, though often unconsciously, is on the side of humanity. However ruthlessly he may be organized, repressed, or ignored, he continues to think, feel, and function as a person. If wounded, he still bleeds; if bereft, he still mourns. He has his life to order and his decisions to make, not less but more because his epoch is suffering from an epidemic of suicide.

Last January I read the following in the correspondence columns of the *Daily Herald*.[5]

> Does this fight for freedom include personal feelings? If so, how is this for an example? My wife's sister has a little girl of six years who has just recovered from a serious operation and illness. The mother's pent-up emotions and worry have resulted in her going down with a bang, which has put her in hospital. Her little girl is left fretting at home with her father. The father (warden since the outbreak of war, who has done a good bit of Blitz work in Islington, where he lives) suddenly becomes de-reserved, gets his calling-up papers, and has to go, after trying to get extension of time. He has a shop – just a small business, newspapers and sweets – and he is not allowed time to do anything about this, his wife, or his child.

I am sure you will agree that the officials in charge of this man's case needed an elementary lesson in psychology. Repressed suffering and ignored anxiety are disruptive forces. They cannot be repressed and ignored for ever – and the longer repression continues, the more dangerous its potential explosiveness becomes.

You have probably heard the cliché which affirms that suffering ennobles individual character. This is often true – at any rate of the finest people – in the perspective of a lifetime, when the memories of past pain confer a compassionate wisdom. But it is seldom true when the suffering occurs. Only saints are capable of early redemption by its means. Amongst the majority, it is apt to crush the weak and embitter the strong. Immediate reactions to it so often take the form of resentment, intolerance, and exasperation, that the sufferer tends to antagonize his friends at the moment when his need of them is greatest. Unless he possesses a power of understanding which is given to few, he is likely to provoke in them the cruel but spontaneous human impulse to belabour the person who is 'down', and thereby to increase his own state of misery.

There is no type of suffering of which this is truer than shame. Suffering takes many forms, but humiliation is characteristic of all but the noblest kinds. If ever you are humiliated by something, or somebody, look into your own heart. You will realize then that the person who points out your faults to you in bitter words – faults, perhaps, which you are slowly perceiving for yourself – is not really doing you, or anybody else, a service. Candour, so called, is seldom the friend of charity or pity.

Thanks to its inhumanity, its disregard of the individual, and the extremity of its policies, modern war causes humiliation on a worldwide scale. The only type of extensive degradation which it temporarily reduces is unemployment. In the place of those whom it rescues from this form of suffering, it creates a number of new and growing categories. These include the victims of racial antagonism, prisoners and internees, refugees and compulsory emigrants, evacuees, the sick, the famished, 'suspect' aliens, conscientious objectors and the pacifist minority to which they belong.

This minority, whose attitude you question, is probably doing as much to defend the power and value of the individual against the repressive tendencies of a totalitarian age as any other group in Britain or America. Its members have also had reason to consider the now universal problem of suffering and humiliation. But of this I shall write more in a later letter.[6] In my next I will tell you something about the men and women who are trying to persuade their contemporaries to renounce war, just as John Brown and his followers once persuaded them to abandon slavery.

Notes

1 November 21st, 1917.
2 By Ronald Hyde in the *Evening Standard*, March 6th, 1942.
3 The *Doncaster Chronicle*.
4 January 7th, 1942.
5 January 24th, 1942.
6 See Letter 4.

3
Numbered with the Transgressors

When a war is in progress, the minorities who resist it can hardly expect to enjoy themselves. Being foremost among the opponents of power-politics, they become inevitably its victims. They are, in fact, fortunate if they escape with their necks.

We do not know but can guess the fate of any person who takes a minority position – such, for instance, as Tolstoy's – in Russia today. His situation would be much the same in Germany or Italy, though most people are unaware that the small Quaker groups in Germany have been allowed to continue under the Nazi regime, owing to German gratitude for the relief work done by the Society of Friends in Central Europe after the last War. On the European continent, the strongest anti-war minorities are probably those in Denmark, Sweden, and Finland. But, in one country or another, the fortunes of pacifists in war-time are likely to run the whole gamut of humiliation, from death at the hands of a shooting squad to ostracism and misrepresentation by their former friends.

Here in Britain, as in the United States, war resistance was first organized between 1914 and 1918. I had no contact with pacifism then, and hardly knew that it existed. If I had known, I should probably have disapproved. There seemed at that time to be no general realization – as there is today, though most people refuse to admit it – that men if they choose can end war by the exercise of their wills. Rather it was regarded as an act of God in one of His bad moods.

The reason was that for exactly a century no important war had touched everyday life in this country. After the Napoleonic Wars ended in 1815, Europe had peace until the Crimean War 40 years later. This affected England very little, except for the work of Florence Nightingale and her subsequent foundation of modern nursing. Nothing happened which could be compared with the American Civil War and its effect on homes and families. During the Boer War – another far-off campaign – a small 'pro-Boer' party emerged, but the mass of the people hardly realized its existence. Apart from the teaching of the Quakers, the pacifist movement in Britain began when the No-Conscription Fellowship was formed in November 1914.

The total membership of this body was quite small. It included about 16,000 conscientious objectors; and of these rather more than 6,000 were arrested during the War. You probably wonder why these men objected to fighting, and why anyone joined a society which brought together thousands of men and women whose views were as varied as their incomes and occupations. It is difficult to define the common factor that united them, but at bottom it was probably the same strong sense of human kinship which came to me through a different experience.

These people refused to kill, or to sanction killing, because to them human life was sacred, and far more important than the conflicting interests of governments. They had, of course, other beliefs, which sprang from this fundamental one. Many of the opinions which they came to hold by the time that the War was over were the result of their resistance rather than its cause.

For many reasons the organization of this group of pacifists was a revolutionary event, and they incurred the unpopularity of most pioneers. A number of conscientious objectors were roughly treated by military tribunals. Some were secretly shipped to France and 30 of them were condemned to death, though the sentence was later commuted. Other pacifists who were above military age suffered penalties too. The most famous of these was Bertrand Russell. Because of his opinions, and the way in which he publicized them, he was deprived of his lectureship at Cambridge, refused an exit permit to deliver a series of mathematical lectures at Harvard, and forbidden to give addresses in English towns within 'prohibited' areas.

Most of those early conscientious objectors were very young; their average age was only 21. Of course they were called 'conchies', 'shirkers', and a dozen other complimentary epithets, but by resisting persecution and threats of capital punishment, most of them managed to win respect from their contemporaries in the Army. There is not as much difference between the outlook of a soldier and that of a conscientious objector as you might suppose. The more thoughtful and intelligent amongst both groups have similar ideas of the ends they want to achieve; what they differ about is how to achieve them.

You may think it a paradox to say that one man who fights and another who refuses to fight have the same purpose in mind; but it is often true. It means that the soldier and the conscientious objector have far more in common than either has with the comfortable type of civilian who incurs no criticism, runs no risk, and makes no sacrifices worth mentioning. I shall probably have more to say on this subject in a later letter.[1]

When the last War ended the No-Conscription Fellowship had fulfilled its purpose, so it was wound up at a 'Concluding Convention' in November 1919. The address on that occasion was given by its President, Clifford Allen, afterwards Lord Allen of Hurtwood. This is what he said about the pacifist movement:

> We believed it to be our duty as a minority still to maintain our position and to preach the gospel of peace in season and out of season. Such a minority will constantly fail; it will be constantly overwhelmed by the tide of warlike passion, but it must stand apart, not idly, but seeking peace. It is clear that such action must be unpopular, and not the cleverest tactician among us can make it acceptable to public opinion in time of war; but as the ages go by that minority will increase, and we must hope that each war will produce a larger and larger number of people in all lands who will desire to create the machinery of peace.

Clifford Allen did not live to see the Second World War, but his hope for an increase in the minority which he helped to organize has been fulfilled. Up to September 6th, 1941, 62,099 men had declared themselves to be conscientious objectors. 'Total war' has also brought a problem which Lord Allen did not foresee, for thanks to the conscription of women we now have women C.O.s as well as men. A number of these have already been registered by the Tribunals.

These conscientious objectors represent only one part of British pacifism; they are the section which happens to be of military age. Behind them stands a growing body of public opinion which is not easy to estimate or locate. Some who secretly agree with the pacifists are afraid to say so, because in totalitarian warfare the power of even a democratic state to punish and penalize 'dissenters' is overwhelming. You know already that though British pacifists have not been invited to sacrifice their necks, they arouse considerable hostility and incur many disadvantages.

On November 25th, 1941, Dr Alfred Salter, M.P. for Bermondsey, made a speech in Parliament which has since been published under the title *Testimony to the Commons*. In this speech he stated that about two million people – that is, one in every twenty of our population – shared the views that he expressed. By no means all these pacifists or near-pacifists belong to societies. But the Central Board for Conscientious Objectors has 17 so-called 'constituent bodies', whereas its predecessor, the No-Conscription Fellowship, had only three.

In the present War the men and women who have become war-resisters and joined pacifist organizations are probably more conscious of their motives than their predecessors in 1914. That does not mean that their reasons are identical. Some have joined from religious, some from political motives; others – probably the majority – from a combination of the two. Quite a number reached their beliefs through suffering and conflict after fighting in the last War. A few, like myself, were brought by their service to prisoners and the wounded to realize the unifying quality of the needs and emotions which humanity shares.

Before I go on to explain these beliefs in greater detail, I want you to understand that it is war itself which has drawn so rigid a line between pacifists and the many other 'men of good will' who do not fully share their convictions but have a great deal in common with their outlook.

During the two decades between the wars, the peace movement in Britain and the United States was not arbitrarily divided into 'pacifists' and 'non-pacifists'. For years many people who are now pacifists supported, as I did, the League of Nations and 'collective security'. Although they regarded the sanctions clauses of the Covenant as mistaken, they felt that the League was the best existing instrument for the organization of peace. At the famous Oxford debate in 1933 on the motion that 'This House will not fight for King and Country', its supporters included believers in economic sanctions and advocates of an international police force. The proposer of the motion, Dr C. E. M. Joad, himself abandoned pacifism in 1939.

It is only within recent memory that the League of Nations, discredited by its failure to prevent wars in Manchuria, Abyssinia, and Spain, degenerated into a mere instrument of French policy for preserving the *status quo* set up at Versailles. After the League had lost its authority, 'collective security' began to mean, in the words of Canon H. R. L. Sheppard, that 'if war came, everybody would be in it'.

When it became evident that international relations were likely to become much worse before they would get better, the pacifists within the peace movement had to choose between a new refusal to compromise, and the illogical position of supporting the 'fight to a finish' which they believed to be the last and worst stage of a series of disastrous government policies whose successive steps they had consistently opposed. I remember some weeks of mental anguish during 1936, when I went about the country making speeches in support of the League of Nations largely in order to test my own allegiance. This experience finally showed me that I agreed less with the position I was trying to maintain, than with the convictions expressed by some of my questioners. These were mostly members of the Peace Pledge Union, which Canon Sheppard had recently founded on the basis of the pledge: 'I renounce war and I will never support or sanction another.' When I found that I shared their views, I had no honest alternative but to join them.

As war came nearer, the members of this and other pacifist organisations realized, from the experiences of war-resisters between 1914 and 1918, that pacifism in war-time would mean ostracism, isolation, misunderstanding, a sense of frustration, and often the loss of salaries and careers. But the great majority, especially among the rank and file, refused to buy security and approval at the cost of integrity. A few, including some of the best known, did change their minds and support the War, and were criticized, often rather bitterly, by their former colleagues for doing so.

Many of those who recanted were undoubtedly sincere in their belief that yet another war must be fought to end war, and that Christian civilization could be defended by cruelty, falsehood, vengeance, and other methods which Christ Himself repudiated. The difficulty of their critics arose from the fact that it is extremely easy to rationalize yourself into supporting a war, especially if you have a dignified reputation or hold a key position, when you know that you will incur official disapproval if you fail to do so. It is always hard for people to believe in your sincerity when your change of opinion coincides with your interests. That is why the only ex-pacifists whose conversion carries conviction are those who join the Forces and thus add to the jeopardy in which total war places their lives. Few, however, do this. The more eminent are normally over military age.

Now I want to go back and try to explain more fully why thousands of people, some for religious and others for political reasons, have felt compelled to renounce war. This is not actually the negative position that it sounds, for you cannot renounce war without making positive endeavours to build up the kind of civilization in which war will have no part.

Most of the religious war-resisters – those, that is, who call themselves 'Christian pacifists' – believe that war arises, not from the evil ambitions of any one man or the inherent wickedness of any one nation, but out of the collective sin of all mankind. To them, war is an inconceivable remedy for the evil from which it springs, since those who make war justify themselves by laying all the blame on the other side, whereas the first and most necessary step in the cure of sin is to acknowledge the extent to which we ourselves are at fault. Because Christ, instead of blaming human failure, took the sins of the world upon Himself and sought to atone for them by His own death, His Christian pacifist disciples believe that 'suffering which may lead even to the Cross' is a high road to redemption.

This type of suffering – unlike the kind of which I wrote you, that embitters because it is resisted and resented – does ennoble the individual, for its secret is a love that can be neither destroyed nor conquered, whatever penalty it may be called upon to bear. 'Not by power, nor by might, but by My spirit, saith the Lord.' This belief is not peculiar to British and American Christians. Recently, at a meeting in London, I heard Pastor Franz Hildebrandt, the German refugee friend of Pastor Niemoller, testify to his conviction that the only way of permanently overcoming the evil of Nazism is the acceptance and pursuit of the way of Christ 'through pain, death, and hell'.

Those pacifists who do not call themselves Christians – though some of them accept the term 'religious' – approach international relations from a political or an economic standpoint, usually backed by sound historical knowledge. Like the Christians, they believe that Hitler and the other dictators are not causes, but consequences, of the evil in the world, and repudiate the hysterical thesis that the Germans are a race of butcher-birds, damned and doomed to make perpetual war upon their neighbours by an exceptionally large share of original sin. Their study of modern Europe shows them that Germany is merely the latest of a long line of 'aggressors' which have included Britain, France, and Spain; and that, like other nations, she is composed of several politically conscious minorities and a large, politically indifferent majority. She has, however, been unlucky in her history, which has tended to encourage the domination of her most aggressive minorities because her great natural energy and efficiency have been accompanied by a lack of certain advantages enjoyed by other leading states.

You will remember that Germany, like Italy, was one of the last European nations to be unified. Until the late nineteenth century, the names of both these countries were merely geographical expressions. Because Germany had been the starting-place of the Reformation, she became, a century later, the battleground of the Wars of Religion. That Thirty Years' War, imposed upon Germany by the policy of Cardinal Richelieu and prolonged by France in order to keep a potential rival disunited, retarded the civilization of Central Europe for over a century. The Treaty of Westphalia, which ended the War in 1648, created a mosaic of over 300 backward little states. When Napoleon formed the Confederation of the Rhine in 1806, he did more to unite Germany than anyone before Bismarck. The Napoleonic Wars finally left the number of the German states reduced to about 50.

Meanwhile, the sixteenth-century nationalism of Britain and other European countries had grown into imperialism. By the second half of the nineteenth century, the British Empire had been built up through a combination of conquest, purchase, and that subtler method known as 'appropriation'. When young Imperial Germany, led by Bismarck who had united her and later by Kaiser Wilhelm II, desired also to join in the race for territory, she found that she could do so only at the expense of her neighbours in Europe.

The same story is approximately true of Japan, with her rapidly increasing population (now nearly 90 million) packed into a very small country. When she too caught the appetite for empire from the West, the choicest Pacific territories were already occupied by the British, French, Dutch, and Americans. The other day I found an old Nelson's Encyclopadia, dating from my schooldays just before the last War, which gave the following figures in square miles for the territories of the chief Colonial powers at that time: United Kingdom, 11,305,126; France, 4,732,100; Belgium, 910,000; Portugal, 803,310; Netherlands, 782,800; United States, 728,330; Italy, 185,200. Germany then owned 1,027,820 square miles, mainly composed of the African territories which she lost by the Treaty of Versailles. Japan, with a population already numbering more than 48 million, possessed only 114,750 square miles. The annual value of the United Kingdom's exports were then £600,000,000; of Germany's £375,000,000; of Japan's only £42,170,000.

After their defeat in 1918, the Germans were weary of war and longed for a period of uninterrupted peace. As I discovered for myself when I visited Germany in 1924, they were in no mood to embark upon a new career of national conquest. According to such eye-witnesses as William L. Shirer, the author of *Berlin Diary*, they were not in that mood even by 1939. But after the victorious Allies had seized German colonies and coalfields, written War Guilt clauses into the Treaty, imposed astronomical Reparations, occupied the Ruhr, admitted Germany with belated clumsiness into the League of Nations, and refused Chancellor Brüning his peaceful economic *anschluss* with Austria, the Weimar Republic was not unnaturally discredited, and democracy associated with humiliation. Hitler came into power because the Nazis appeared to be the only alternative to perpetual dishonour and depression.

During those years between the wars, Britain and the Dominions, by means of the Ottawa Agreements and other measures, had also put up such high tariffs against German exports that Germany was short of foreign markets. Japan had to face not only tariffs but American and Dominion immigration laws, which prevented the nationals of her overcrowded islands from emigrating to large, under-populated, and convenient territories. (The population of Japan proper averages 469 to the square mile. In the U.S.A. the average is 41; in Canada 3; in Northern Australia under 2.) This policy strengthened the militarist minority in Japan, which is no more typical of the whole people than the Nazis are of Germany. Only recently the Japanese Christians presented 7,000 yen to the Christians of China for rebuilding their bombed churches.

What, you may ask, would pacifists have done about the German Nazis and the Japanese militarists? The answer is that they would probably never have had to deal with them. If, by some unlikely miracle, anti-war reaction had put a pacifist British government into power in 1918, the oppressive clauses of the Versailles Treaty would never have been written, nor the Ruhr occupied; liberal Germany would have been admitted immediately, and cordially, into the League of Nations, and Stresemann, the chief representative of the Weimar Republic, would have had much more than the 'one important concession' for which he vainly asked in order to 'save peace for this generation'.

Other opportunities for constructive reconciliation would also have occurred. In 1932 the Disarmament Conference assembled at Geneva.

Six years later, in February 1938, Dr Hugh Dalton, until lately the Minister of Economic Warfare and now President of the Board of Trade, recorded of this gathering:[2]

> On February 10th, in the first debate of the conference, Italy proposed the abolition of all bombing aeroplanes. Germany, Russia, and other states supported. The United States of America was friendly to the idea, and in June President Hoover definitely came out in favour. From the first Sir John Simon and Lord Londonderry resisted and obstructed.

The representatives of a pacifist government would not have 'resisted and obstructed'. They would have encouraged Italy, Germany, Russia, and America – just as they would have supported the earnest plea for the entire abolition of bombing made by Japan before an official commission of jurists at The Hague in 1923, which was rejected owing to French and British opposition.[3] They would not have allowed the World Economic Conference of 1933 to become a failure owing to the operation of those sinister underground forces known as 'interests'. Nor would they have permitted the Report issued in 1938 by M. van Zeeland, the Belgian Minister who was asked to draw up a plan of international economic reform, to be quietly pigeonholed thanks to those self-same influences. They would have used that Report as the basis of an international New Deal, by which the needs of the 'Have-Nots' would have been met by the 'Haves'.

It hardly seems open to doubt that a generous policy of this kind, actuated by the Christian principle of do-as-you-would-be-done-by, would have been better for Britain, America, Russia, and China (to say nothing of menaced and mishandled India) than the present enormous and suicidal War. Even some of the 'interests' themselves would have suffered less. The European owners, for example, of tin mines, oil wells, and rubber plantations in Malaya and Java would not, by sharing their profits, have lost their possessions so completely as they have lost them now.

Yes, you may say, but surely it was no good contemplating a policy of this kind once the Nazis were in power, and much less after war had begun? Pacifists, both Christian and political, would reply that the way to prevent Germany (or Japan) from pursuing a career of conquest is not, yet again, to smash, humiliate, and punish – if you can – the aggressive minority and the whole country in its name, with the inevitable consequence of a new desire for revenge and a new growth of militarism. It is to divert the course of history into new channels.[4]

Perhaps you will argue that you cannot do this unless you first beat the Nazis by their own type of weapon? But the evidence of the last War, and the dictated peace which followed it, does not suggest that the demonism rampant in the world today – and not only in Germany and Japan – can be permanently conquered by fire and steel. At best it would only be temporarily subdued, to come back stronger than ever. Nazism thrives, as we see repeatedly, on every policy which provokes resistance, such as bombing, blockade, and threats of 'retribution'. These measures unite the despairing German people behind the oppressor within, as the only means of withstanding the enemy without.

Supposing instead that we were to offer, by every broadcasting device available, an immediate armistice or 'peoples' peace', coupled with food for the starving, suggestions for reciprocal disarmament, and generous undertakings to share the rich resources still within our control? This might not influence the German and Japanese militarists, whose power flourishes on Allied ruthlessness, but it would remove from the German and Japanese people their main reason for supporting their present leaders. A great magnanimous gesture is not merely one way of ending a war; it is the *only* way to end it without sowing the seeds of another conflict.

There is not, of course, the slightest hope of obtaining such a policy from the present governments of the warring democracies. Should pacifists, as some people argue, therefore cease to demand it, and instead endorse the continued frenzy of destruction which all history shows to be the course least likely to lead to permanent peace? Are we to say that our own policy, which has never been tried, is a failure because the *opposite* of it has led the world to disaster?

In the name of truth we cannot say so – any more than the despised pioneers of the early Church could cease to urge the standards of the Kingdom of Heaven because these standards were unlikely to be achieved under the Roman Empire. We are still very far from that New Jerusalem; but if those struggling and imperfect early Christians had abandoned its ideals as 'impracticable' in the same way as modern pacifists are urged to repudiate their conception of a true international society as 'utopian', the teaching of the Gospels would not have survived through the ages to be a constant summons to courage and a perpetual challenge to despair. So long as a generous and imaginative policy remains untried, we cannot support a war which we believe to be caused by the failure to try it.

Notes

1 See Letter 5.
2 See *Arms – and the Men*, by Oliver Brown (published by R. Thomson, 11 Spoutmouth, Glasgow, C.3).
3 Tokyo Correspondent of *The Times*, October 16th, 1937 (quoted by Oliver Brown).
4 See Letter 10.

4
Humiliation with Honour

The comparison in my last letter of modern pacifists with early Christians was not due to a belief that minorities as such have any peculiar claim to merit. It was suggested by a note in the Chicago magazine *Unity*,[1] which compared pacifism in war-waging Britain with the position of the pioneers who founded their Church in a totalitarian world dominated by the most comprehensive military power known to history. The writer said:

> There was no answer to these men and women who refused to sacrifice to the Emperor and to take up arms to fight under his banners, when they talked about the God who 'hath made of one blood all nations of men' – no answer except persecution. But persecution has not come in England – to the glory of England be it said.

It is true that the British governments of the Second World War have learned from the records of the First that persecution tends to confer an inconvenient crown of glory upon its victim. No such radiant compensation attaches to a quiet and gradual process of suppression, to an undermining of prestige and a steady denial of opportunities, nor even to the unpublicized monotony of life in Holloway or Wormwood Scrubs. Persecution on the Nazi and Soviet models has therefore been avoided and a reasonable measure of free speech maintained. Yet whenever danger has threatened and panic awakened, the humiliation of pacifists has provided a safety-valve for hysteria and intolerance. Their treatment, particularly during the summer of 1940, has usually been an accurate thermometer by which to register the national temperature.

The opening 'Sitzkrieg' period of the War was one of marked toleration towards all minorities. Four months before it broke out, on May 4, 1939, Mr Neville Chamberlain made a pronouncement defining the official attitude towards pacifists. He said:

> We all recognize that there are people who have perfectly genuine and very deep-seated scruples on the subject of military service, and even if we do not agree with those scruples, at any rate we can respect

> them, if they are honestly held . . . I want to make it clear here that in the view of the Government, where scruples are conscientiously held, we desire that they shall be respected and that there should be no persecution of those who hold them.

This statement meant that, between September 1939 and June 1940, the majority of pacifists kept their posts and were able to say and write more or less what they chose. The few who had professional or business engagements abroad were not only allowed but, in one or two cases, actually encouraged to fulfil them. But with the fall of France a violent and unfavourable change of policy swept everyone who could not support the War into the ranks of the transgressors.

Throughout Britain during that ominous summer, Government and people were seized by a panic of suspicion. The national pursuit of spies and Quislings reached a point at which loyal pacifists were labelled 'Fifth Columnists', and were watched with a consternation comparable only to that which led to the reckless imprisonment of Gandhi and the Indian Congress leaders in August 1942. At the same time the House of Commons, frantically apprehensive of invasion and other vaguer threats of which it was even more terrified, made over to well-meaning but limited men alarming powers of arbitrary legislation which only the highest type of spiritual genius is fit to exercise.

If the pacifist victims of these emergency powers had been German instead of British, their position – though politically more dangerous and physically far more painful – would have been more comprehensible to the authorities. They would have been ranked with the Catholic Bishops of Münster and Berlin, the Protestant Bishop of Württemberg, and the imprisoned pastors of the Confessional Church, in their 'mental fight' against a regime which lives by war and oppression. That would be the role of British pacifists if ever this country were to be conquered by Fascism – either from within or from without.

But in 1940 it was easier for prejudiced or confused mentalities to identify the opponents of war with the supporters of Hitler, than to acknowledge their endeavours to live by the standards of a society which has not yet come. Under revived or newly created regulations, pacifists began to be arrested for 'insulting words' or 'spreading alarm and despondency' – though most of them were much less alarmed than the Government. Six leading members of the Peace Pledge Union were prosecuted for the exhibition of a two-year-old poster, and were tried and fined at Bow Street Police Court. The very supporters of pacifism who earlier in the War had been encouraged to travel owing to their international contacts, were now refused permission to leave the country, and have never since recovered their right to fulfil professional engagements abroad.

When the danger of invasion decreased this phase of political panic passed, leaving many who had suffered from it a little ashamed of their fears and behaviour. But the arbitrary authority given to the Government proved easier to confer than to repeal. By permitting it to draw up whatever regulations 'appear to His Majesty to be necessary or expedient', Parliament had destroyed the check by which the Courts were obliged to declare a regulation invalid if it went outside the powers given by the Act.

It was thus that the threat to the freedom of pacifists, like that of other minorities, became permanent and was sometimes carried out. More than a year after the fall of France, one of the best known members of the Fellowship of Reconciliation, Muriel Lester, was taken off an American ship at Trinidad after two years of lecturing in the United States. She was interned there for four months, deported to Britain, detained in a Glasgow police cell, and imprisoned for several days in Holloway Gaol, before she recovered her freedom. Yet no charge was ever made against her, and she was given no opportunity of self-defence.

In a sense, of course, she was, and is, in disgrace. So are many other once respectable people, including your mother. I want, therefore, to try to show you that this kind of disgrace is not really disgraceful. I hope to make you understand that even the humiliations which it involves can become assets which add to knowledge and increase spiritual power if they are regarded as valuable experience obtainable in no other fashion.

In my last letter I endeavoured to explain the pacifist belief that Christ's injunction to 'overcome evil with good' lays upon His followers the obligation to pursue a way of life which uses, like Gandhi in India, only the weapons of the spirit against the powers of darkness, and directs those weapons first against sin in themselves. It also obliges them to affirm that pursuit publicly, and to accept, with as much charity and as little bitterness as possible, whatever penalty their faith may involve. A decision of this kind has been a turning-point in the lives of many people. In the deliberate, unresentful choice of avoidable punishment, we come as near as most of us ever get to the heart of religious experience. Though the results of enforced pain may be evil, the redemptive function of pain consciously chosen and accepted seldom fails.

But redemption does not happen automatically. It comes, as Olive Schreiner said, from within. 'It is wrought out by the soul itself, with suffering and through time.' It represents the reward of victory in a conflict between man's impulses and his will.

Just because they are helpless against the Juggernaut power of the totalitarian machine, the victims of war, as my second letter indicated, become sources of potential disruptiveness. This is a dangerous threat not only to the present but to the future of any society in which they are found. I am not now referring only to members of minorities, but to the millions throughout the world who suffer because a few conspicuous men play the game of power-politics. Since many years will certainly pass before the statesmen of a more enlightened era quit power-politics for welfare-politics, the thwarted energies of these multitudes, unless consciously self-disciplined, are likely to find a destructive outlet.

Nazism and Fascism themselves are typical results of resentment and humiliation. That is why I keep on insisting that you cannot defeat an armed doctrine by bombs and tanks. All you can do is temporarily conquer the men and women enslaved to this doctrine. If you kill them the idea passes to others, rising like a phoenix from their ashes. You may thrust a doctrine underground by victorious might, but you will not destroy it. There is only one way of getting rid of a bad idea, and that is to replace it by a better. Evil arising from injustice will never die until the injustice has been supplanted by justice.

Sometimes you will find unassertive individuals who are merely driven by conquest or oppression into the death-in-life of an enervating defeatism. This, though fatal to themselves, is largely negative in its effect on society. But the more affirmative type will always be tempted to respond with the pugnacity which develops in isolation and the vindictiveness which shapes itself into plans for revenge. These plans will be directed against whatever, or whomever, has precipitated the humiliating situation – a state, a ministry, a political party, a person. The resentment which inspires them may be visited upon the Government officials who have applied emergency regulations against the sufferer in a fashion which seems to him deliberately and personally inhuman, rather than an expression of the preoccupied indifference which is usually responsible. Or it may be reserved for other more fortunate individuals who have found excellent reasons for pursuing a less thorny path, thereby avoiding frustration and the bitterness which it causes.

You can take it that every member of an unpopular minority is liable to the growth of a persecution mania. I have had letters from young women who lament the deliberate malice of their former friends, and from young men who believe that the authorities are trying to starve or terrorize them into military service despite a recognized conscientious objection. Some pacifist groups behave like members of a leper colony; they embrace persecution and misrepresentation instead of boldly disregarding them. Thus, by self-isolation, they add to the ostracism which is only too convenient for war-makers to impose upon critics who insist on exposing hypocrisy and telling the truth.

In this way humiliation may lead to a pathological condition in which, as in a diseased body, the germs flourish of still worse conditions – permanent resentment, ingrowing hatred, anti-social conspiracy. The whole history of post-war Germany illustrates this process. Yet persons enduring persecution are actually rendered capable, by that very fact, of becoming more sensitive to the vast accumulation of pain in the world, and thus of learning a compassion for others worse off than themselves in which their own sorrows will be speedily disregarded.

Let us imagine two typical men whose minority opinions have brought them humiliation. One is a public figure – a politician, perhaps, or a writer. Because his country has gone to war, suffered reverses, and become panic-stricken, he suddenly finds himself 'suspect' and unpopular. He is forbidden to travel to other countries where much of his work was done, refused permission to lecture or attend conferences abroad, compulsorily separated from old friends or members of his family living overseas, and even prohibited from entering 'defence' areas of his own country in order to obtain material for books or speeches. He is, in fact, in the same situation as that which prevents me from coming to America and seeing you.

There are few sources of frustration more bitter than the knowledge that some rare gift, some skill perfected by the practice of half a lifetime, is wasting unused; denied not only its normal expression, but prevented from reaching those who would benefit by it for reasons which have no direct relevance to their need. The note of grief for such wastage echoes through Milton's most famous sonnet: 'And that one talent, which is death to hide, Lodged with me useless . . .'

Nobody likes, in addition, having his name on a 'list', or knowing that his private letters are carefully read and copied by strangers and then filed in a government 'dossier' for inspection by officials. It involves a loss of freedom which is none the less galling because it is not obvious. So though the blow to our public man's pocket is probably severe, it is nothing to the assault on his pride, skill, prestige, and affections. He had assumed his integrity to be self-evident, and had never imagined that a deep spiritual or intellectual conviction – arising, perhaps, from his own experiences in the last War – could be interpreted as a readiness to betray his country.

The other man, though less conspicuous, has held an equally honourable position. He is a teacher in a secondary school, popular and respected. Suddenly danger threatens his district, and the term 'Fifth Columnist' is hurled at him. His local authority subjects him to humiliating inquiries; his few remaining friends advise resignation. The parents who once asked for his counsel look askance at him; even his pupils, following the example of their elders, cut him in the street. His livelihood and his reputation alike are gone.

Both these men, normally loyal and benevolent, are liable to become disruptive forces. Their bitterness against the Nazis, who may have precipitated the War but are unknown and remote, is nothing to their resentment against the Home Secretary or the local education officer, who are the direct causes of their suffering. They forget to contrast their relative immunity with the torture, imprisonment, and 'liquidation' reserved for minorities in totalitarian states. So they secretly meditate on picturesque revenges, and their hearts turn sour within them.

But we need not leave them there. If either has the capacity for salvation which is inherent in many, he may one day recognize that he has passed through a spiritual experience. Nobody can foresee how this will happen. The winds of God blow as they will, and this man may find himself at his spiritual crossroads through reading a poem, or looking at a beautiful scene, or walking alone in an empty valley on a fine autumn day. But he cannot fail to realize his sudden reconciliation with society, for it will spring from the knowledge that he can only work effectively for the sorrowful and oppressed if he gives up all hope of remaining 'respectable'. Like Christ before him, he must put himself among the felons and endure as best he can the calumnious assertions that are made about him. By descending to the level of the outcasts whom once, perhaps, he contemptuously pitied, he will acquire a compassionate understanding of degradation and its effect upon the mind and spirit.

You remember Winifred Holtby, of course, and her novel *South Riding*. If you have read it, you will recollect the advice which Alderman Mrs Beddows gives to the school teacher, Sarah Burton, in rescuing her from despair after the death of Robert Carne:

> When there's no hope and no remedy, then you can begin to learn and to teach what you've learned. The strongest things in life are without triumph. The costliest things you buy are those for which you can't even pay yourself. It's only when you're in debt and a pauper, when you have nothing, not even the pride of sorrow, that you begin to understand a little.

As soon as you find yourself on a path which can lead to no worldly advantage, you are freed at last from the competitive scheming of seekers after office or place. Once you are in the position reached by this man, you have no longer any share in the perpetual anxieties of those to whom this War appears as a source of power or financial advancement if only the right strings can be pulled, the right support elicited before someone else obtains it. Your road to salvation lies through pain and dishonour, for which there is no competition.

That is why a German pastor detained in prison by the Nazis wrote to his wife: 'It is just this humiliation which is so necessary for us, and God uses it to unlock us inwardly and to ripen us for His Word and for His consolation.'[2] He knew that grievances vanish and vindictiveness disappears with the discovery that every sorrow bears its own compensation which enlarges the scope of human mercy. It is not easy to relinquish self-pity, for this involves the unflattering admission that we have indulged in it. But once we can achieve the imaginative realization that the sufferings of many are greater than ours, we find ourselves possessed with a desire to relieve them which causes our own sense of injury to be set aside and finally forgotten.

You need not imagine that frustration is peculiar to one minority, or even to minorities in general. It is part of the burden borne today by every victim of power in a world where power-politics have reached their nemesis. I have tried to show you that the best hope for all such victims lies in their frank recognition of their danger to their fellows, and in their continued resistance to its growth. They should not be over-troubled if they can manage no more than self-conquest, for that in itself is part of the conquest of its own evil and inertia by society.

Once they have understood that even degradation can be a salutary and creative experience, the country to which they belong is on its way to learning the same lesson. Only humiliation with honour – the honour of self-discipline and of new wisdom wrought out of bitter experience – can save men and women degraded by war from becoming sources of hatred and vengeance, and enable them to contribute in their unique fashion to those abiding things 'which belong to our peace'.

Notes

1 February 1942.
2 See *I Was In Prison*. Letters from German Pastors. (Student Christian Movement Press.)

5
The Functions of a Minority

The many logical arguments put forward in support of war do not alter the fact that it is now the deadliest disease of our civilization, and must be overcome if that civilization is to continue. Mere victory over Hitler will not overcome it. It will only serve to aggravate the disease if the wrongs from which Hitlerism sprang – monopoly capitalism, imperialistic nationalism, poverty, hunger, unemployment, repression – continue unchecked into the post-war era. There is little evidence as yet that those who are conducting the War on behalf of the United Nations propose to check them.

Pacifists are people who want to fight the disease, instead of wasting life, time, and energy in attacking the symptoms. They have come by different roads – the best, perhaps, being actual experience of warfare – to the realization that modern war never achieves even the ends for which it is ostensibly waged, let alone a stable world and a peaceful society. Believing this, they would be false to themselves and their faith in man's capacity for redemption if they supported the present War and collaborated with its leaders.

Perhaps you will feel that those who share this conviction and act upon it ought to be specially enlightened people. A few indeed have been; it can justly be said of such men as H. R. L. Sheppard, George Lansbury, and Max Plowman that they were saints on earth. But unfortunately all three of them are dead; and we, their successors, have to struggle on as best we can without their special gifts of religious insight to help us. Often the words which St Paul wrote to the Corinthians in the early Church are only too true of modern pacifists: 'For ye are yet carnal: for whereas there is among you envying, and strife, and divisions, are ye not carnal, and walk as men?'

Apart from such conflicts – which are, I fear, characteristic of all sinners who are trying to be saints with no better spiritual equipment than other people – there are one or two groups which do the cause of pacifism harm rather than good. I am not speaking here of the few actual shirkers who always manage to creep in war-time under the pacifist umbrella. The movement itself objects to them at least as much as the general public, for they cannot be counted on either to carry conviction before a Tribunal, or to lend a hand in a blitz. They are as disadvantageous to true pacifism as those temporary adherents who wax eloquent in the cause of peace only during the intervals between wars. The two categories about which I meant to write you are the belligerent pacifists, and the self-righteous.

The 'pacifism' of the belligerents is nothing other than a form of inverted militarism. They are incurable minoritarians with a passion for unpopularity, who will make use of any movement which enables them to express their deep dislike of all majorities. Far from attempting to act as reconciling influences, their purpose is usually to challenge and provoke. If they do not succeed in getting themselves persecuted, suppressed, or imprisoned, they are disappointed. They feel that the sole test of their sincerity is the extent to which they can embarrass the Government.

Pacifists of this type forget that embarrassing the Government, necessary as it may be at times, is hardly a constructive occupation. There are many important subjects, such as the state of the post-war world, which are vital to all persons of goodwill and intelligence but in which governments take little interest. You will find that the worse a government is, the less interest it takes.

Apparently it does not occur to these militant pacifists that the most effective method of war-resistance is to increase the number of war-resisters – who can only be drawn from the ranks of the war-supporters, and are not attracted by provocation. The best method of making breaches in the solid wall of war-acceptance is the joint consideration by pacifists and non-pacifists of problems in which the evil consequences of war are most apparent. It is clear, for instance, that in the bombing of civilians, and the food blockade of Europe, there is a departure from the standards of international law never equalled in previous wars.

The pacifist's contention that modern war, with its huge apparatus of propaganda, involves a process of spiritual deterioration which destroys such moral purpose as it ever had, is here supported by obvious facts which no reasonable person can deny.[1] But belligerent pacifists regard the making of converts by an assemblage of facts and an appeal to reason, as 'being respectable', or 'currying favour with the public'. They do infinite harm to pacifism by giving well-disposed people the impression that the movement is solely composed of irrational, obstinate, and irresponsible fanatics.

With the self-righteous we can have more sympathy, for it is always difficult to put a deep conviction into words without sounding 'smug'. I do not criticize them because they express themselves clumsily or priggishly; even a practised writer of many years' standing finds it hard to say precisely what he or she means on a matter of conscience. Where they fail is in their tendency to regard fighting men, and those who accept military methods, as thoughtless and ignorant, and to forget that the young soldiers, sailors, and airmen of today are as much victims of power-politics as themselves.

Nobody could make this mistake who had once been part, even as a non-combatant, of an army at war, and had seen the sacrificial spirit in which young men with their lives unfulfilled endure their disabling wounds or go out to die. Ever since I watched the British soldiers at Etaples talking to my wounded German patients, I have realized that the fighting volunteer and the pacifist have far more in common than either has with the 'old men' who send the one to death and the other to prison.[2] I know one young soldier, determined never to kill a fellow creature, who has become a volunteer member of a bomb-disposal unit because he suspects that in his case a 'conscientious objection' would be mainly a rationalization of fear and distaste. The test in each instance is the readiness to give up everything, including life itself, for the sake of a moral conviction.

We must recognize that, in the present state of opinion in this country and the United States, the decision to fight was inevitable for most of those who made it. A friend of the late Evelyn Underhill, Mrs Marjorie Vernon, emphasized this point in a letter sent to the Anglican Pacifist Fellowship. She wrote:

> God, we are taught, judges us by our fundamental 'intention'. It seems to me that the average Englishman's 'intention' in taking part in this war is to vindicate freedom, truth, mercy, and justice. What would his 'intention' be if he kept out of it? Remember, England is not a truly Christian country. Ethically, perhaps she may lay claim to be – though even here there are gaps – but on the religious, the supernatural level, she certainly is not . . . If all England were profoundly Christian her refusal to fight would not be a negative refusal, but would be vigorous and positive; not so much that she would not fight as that she *would* do something else. Realistically believing that 'to them that love God all things work together for good', she would be content to follow Christ's teaching, leaving the outcome – whatever it might be – in God's hands and fully accepting it . . . At present,

> however, such an attitude and such behaviour is practically unthinkable. Only a few people born out of due season see these truths, which belong to the future rather than to 1941.

Despite the fairness of this judgement, I receive many letters from men and women in the Forces who want to discuss the justification and purposes of the War. Later I shall have more to say about the bond of sympathy between the convinced pacifist, of whatever age, and the young generation which is again called upon to offer up its health, strength, and opportunities in a war made by elderly politicians whose aims and standards it does not share.[3] But the development of that bond of sympathy depends upon our ability, as pacifists, to use rightly the type of sacrifice which we are called upon to make. It is nothing, in any case, to the sacrifice demanded of the soldier. If he can face wounds and death, whether endured or inflicted, we can surely learn wisdom from criticism and degradation.

The essence of such wisdom is a clear understanding of pacifism and its function. Some people join the pacifist movement without examining either the ethics or the politics of their position, which explains the superficial escapism of inverted belligerency and government-baiting. It is the pacifist's obligation to be both realist and idealist; to face existing facts while never losing sight of the world which he desires to create. His part, as a living leaven within the lump of popular traditions and assumptions, may seem trivial in itself, yet his task is nothing less than an attempt to change the thinking of his nation, and beyond that of a greater society.

He begins modestly by trying to enlarge – sometimes only one by one – the circle of those who endeavour to shape their lives in accordance with a particular set of values. By these 'values' I mean the conduct of that ideal community to which poets and prophets have given many names. Christ called it the Kingdom of Heaven. For Dante and Milton, it was Paradise; for Sir Thomas More, Utopia; for Blake a mystic 'Jerusalem' to be built by 'mental fight'.

The movement that seeks to create this community which knows neither force nor frontiers is inevitably a revolutionary movement. It is a society within society, a living force which depends neither on economic systems nor political machinery – though it may work through both – but upon the power of the spirit. Today it is the only movement which possesses this revolutionary character. The once progressive 'Left' has become reactionary; it is in alliance with those forces which applaud totalitarianism and war. It merely prefers one form of totalitarianism to another. Hence the paradox by which gatherings called to reaffirm the democratic principles of free speech and a free Press are crowded by the supporters of a regime which suppresses both.

Perhaps you feel that a minority so small, surrounded by powerful forces so adverse to its growth, is unlikely ever to achieve its purpose of leavening the lump. But you must remember that nearly all the great revolutions of history not only started as minority movements, but seldom became anything else even when they had succeeded. In the words of John Wesley:

> Give me 100 men who fear nothing but God, hate nothing but sin, and have the love of Jesus in their hearts; and with them I can move the world.

I can think of few important movements for reform in which success was won by any method other than that of an energetic minority presenting the indifferent majority with a *fait accompli*, which was then accepted. The only exceptions are perhaps the great religious movements which, like Christianity, started from a tiny group, but went on to capture half mankind. But even here, as I have tried to show you, there is a wide distinction between the genuine working believers and the vast bulk of lip-servers who make up the substantial majority of both religious bodies and political parties.

Shortly before this War, when we were driving together to a meeting, George Lansbury told me that in his long life he had seen so many apparently hopeless causes succeed, that his faith in the ultimate rejection of war by civilized men and women remained unshaken. His life spanned a period which not only included such movements as slave emancipation, the abolition of child labour, prison reform, trade unionism, socialism, and universal education, but which saw social habits now universally accepted pass from the stage where they were regarded with horrified disapproval to one in which it was forgotten that their desirability had ever been questioned.

Once, on a public platform, I heard Dr Marie Stopes read aloud from a mid-Victorian newspaper a paragraph imputing recklessness and immorality to a certain group of 'advanced' persons. Was this a diatribe against free love, birth control, or companionate marriage? Not at all. The protest was directed against those who had adopted the habit of taking a regular bath.

One of the reforms most successfully carried through by a small minority was woman suffrage. There has been some confusion on this issue, since the first Great War broke out before the effect of the suffragist propaganda was fully apparent. The gap between the demand for female suffrage, and the first partial acceptance of it in Britain in 1918, enabled the opponents of the feminist campaign to say that the vote was conceded to women as a reward for their part in winning the War. But in some countries, such as the United States, where woman suffrage was also granted at the end of the War, the amount of war service performed by women had been very small. The vote would never have been conceded had not a hardworking and articulate minority brought their claims, before the War, to a point where these appeared not only conceivable but rational and just.

In the same way, the movement to abolish war is likely not only to begin, but also to end, as the achievement of a minority which ultimately persuades the majority to adopt its view. The elimination of war and the building of permanent peace differ from other great movements for the liberation of humanity from the evil within itself, only because the sphere of action is so much wider and the problems involved far more complicated.

The minority working for international peace has, as I see it, a fourfold function. Its members have first to fulfil the duty of self-conquest described in my last letter. This really involves a new conception of honour. It means understanding that we cannot exercise compassion until we have endured humiliation, nor effectively help the victims of society until we have been in the dock and the prison beside them. When a man has conquered his own bitterness and learned to wrest honour from shame, he has brought humanity's struggle to overcome war a little nearer to victory. He has also achieved a new kind of freedom which compensates for his earlier frustration.

The respectable have so much to sacrifice: their prestige, their popularity with their neighbours, the good opinion of their friends, the approval of the Government. Naturally they have to think carefully before associating with the oppressed or endorsing 'extreme' opinions, and we cannot be surprised that their decision regarding such association is frequently in the negative. Hence it is a great advantage to be so 'disrespectable' that one has nothing to lose. One can then quit the upper-dogs and, joining the underdogs, begin to see their situation as they see it themselves. Charity born of shared experience is the only power which can finally vanquish evil; that is why those who believe in its authority must take the risks that it demands in a world which does not yet accept it. A Quaker friend once wrote to me of these risks:

> Solid, dogged, unchanging love. Suffering and injustice lovingly endured. Force can restrain the evil man; only love can change him, and when Christ went to the Cross it was no mere acceptance of the inevitable, but His deliberate response to the hatred and selfishness of men, his final assault on the powers of evil. The battle continues now, in every individual life and in a world at war; the spiritual forces we cannot reckon, the human one is the scattered company of all who try to live that way.

Of the three other functions of our minority, I shall write you in my remaining letters. The first, which depends for its value upon the conception that I have called 'humiliation with honour', is to give whatever assistance we can to those victims of power who endure more than ourselves – prisoners, refugees, the starving, the young, the bereaved. We must keep ourselves imaginatively conscious of the cost of war in human suffering; we must also investigate and expose that cost, and find, if we can, some form of redemptive consolation for those who have to pay it.

Secondly, I believe that we are called upon, humbly and without self-righteousness, to keep alive those civilized values of charity, compassion, and truth to which men return with relief and remorse as soon as war is over, wondering how ever their spiritual focus became so distorted. If this includes the duty of protest against hatred, cruelty, self-interest, and falsehood whenever we come across them, we must not shrink from the resulting unpopularity. It is quite different from the unpopularity-for-its-own-sake which the pacifist who would not be happy if he were not victimized sets out to incur. Our business is to keep our heads, and try to see the events of the War in the largest possible perspective against their background of history. We must endeavour to find out and tell the truth, facing facts as they are rather than as propaganda and wishful thinking desire them to be.

Finally, from our study of history and our careful watch on present developments, we must seize upon such evidence of the shape of the future as we are able to acquire. This does not mean constructing an Eldorado of gilt-edged 'Peace Aims' which have no relation to the facts as we know them. It does mean trying, within the limits of our knowledge, to plan for the post-war world as its dim outlines emerge through the stifling smoke-clouds of war.

Our task is that of trying to change the course of history by acting as a revolutionary leaven within society. But we must know in what direction we wish to divert it before we can even begin. Idealists in general, and pacifists in particular, are apt to talk about 'the future' as if futurity were an end in itself. Obviously the future will be no better than the present unless we can learn from the mistakes of the past what its pattern should be.

Notes

1 See Letter 9.

2 The late Colonel T. E. Lawrence bore witness to this resemblance in a famous passage from *Seven Pillars of Wisdom*:

> When we achieved and the new world dawned, the old men came out again and took from us our victory, and remade it in the likeness of the former world they knew. Youth could win, but had not learned to keep, and was pitiably weak against age. We stammered that we had worked for a new heaven and a new earth, and they thanked us kindly and made their peace.

3 See Letter 7.

6
Their Name Is Legion

It is often said by those who support the present War that, of course, they desire peace as much as anyone else, and in any *normal* war they would be pacifists themselves. But this War, they maintain, is different. It was made by international gangsters upon their peace-loving neighbours, and we have no alternative but to punish them and make them pay for their wickedness.

Unfortunately for this argument, the two Great Wars inflicted, for the first time in history, upon one generation have come so close together that many people can remember exactly the same contention being used between 1914 and 1918. In those days it would indeed have surprised me and my contemporaries if some prophet had told us that the time would come when Kaiser Wilhelm II, who was represented as twin brother to the Spirit of Carnage, would be regarded as a gentlemanly and chivalrous opponent compared with that unspeakable mountebank, Adolf Hitler. If you could transport yourself back through the numerous wars of modern history, I suspect you would find that every war is an exception, a *different* war, to those who uphold it.

Despite the strong emotions of my friends who insist that gangsters must be punished, I have never yet obtained a satisfactory reply to my inquiry as to how a war is to be waged which actually penalizes those who made it, while avoiding the rest. Although it would not solve the underlying problems of poverty and unemployment, or avert the envious eyes of 'Have-Not' Powers from rich territories occupied by Britain not for 2 years but for 200, there would be some justice in a war in which the sole participants and the only casualties were the responsible leaders of the countries involved.

But the 'punishment' meted out by war does not fall upon its organizers. Their near relatives seldom run exceptional risks either. It is the 'little man' and his family who are expected to do all the killing and the dying. The price of this War is paid by its innumerable obscure and helpless victims, who are bombed, blasted, massacred, starved, transported, and conscripted without even understanding what it is all about. It is upon the prisoners, the exiles, the homeless, the famished, the bereaved who mourn, and the young who are home-sick and frustrated, that 'retribution' descends in total war. So many men and women the world over now drag out a weary existence behind the bars of prisons or internment camps, that we who are still free to stand in the sun and walk through the fields must consider ourselves fortunate, whatever the War may have cost us.

The other day, in a pamphlet called *A Peace Aims Declaration* issued by the War Resisters' International, I read these words:

> The proper concern of all people of goodwill should be all forms of human oppression, degradation, and unnecessary suffering. War is only one aspect of these evils and it will be found that when these are dealt with the problem of war and peace will be largely dealt with in the process.

That is why I believe that pacifists have a special responsibility towards all those who, like themselves, are victims of power. It is their business to find out what more can be done for these sufferers, not only by others, but by the actual victims in terms of self-conquest and spiritual discipline. The greater the degree of freedom possessed by the pacifist minorities concerned in this quest, the larger the measure of their obligation. For this reason those who are citizens of Britain or the United States, where some considerable remnants of democracy still protect minorities against the full ruthlessness of totalitarianism, are specially called upon to help their fellow victims, and to defend to the full limits of their restricted powers the surviving individual rights in their own societies.

Once this responsibility is recognized, the pacifist's sense of isolation and futility is over. There can be no better contribution to a real victory (which means a victory over the reactionary values of the Hitlerism that we are fighting in every war-making society) than the attempt, however limited in scope, to rescue forsaken individuals from spiritual darkness and physical pain. To help them to find in defeat or exile a new and deeper level of experience is itself part of the endeavour to maintain those religious and social values by which democracy has lived.

Where, you may ask, shall we find prisoners and internees in England? Perhaps it will surprise you to know that, at the beginning of 1942, over 6,500 people were still interned. This figure does not include the detainees in Canada and other parts of the British Empire, where the periodic imprisonment of India's would-be liberators is a strange and tragic comment on Britain's fight for freedom.[1] For one conspicuous internee who has been released, such as Benjamin Greene, the Hertfordshire J.P., a hundred others are detained of whom no one has heard. At the same date there were 519 conscientious objectors in prison, of whom 119 were serving sentences of 12 months. Each new military disaster which lengthens the War, and divides me from you and S., lengthens also the sense of wasted life and power which oppresses these men and women in gaol.

There are also numbers of German and Italian prisoners of war in this country, though military regulations and the hostility of public opinion make it more difficult to give help to them. In a letter to *The Times*[2] I read that some Italian prisoners of war in a small camp in England performed a typical act of courtesy on All Saints' Day, 1941. It happened that this day was also their pay-day, when they received their small 'token' money for work in the fields. The first man called explained that he did not desire to draw his pay. When he was asked what he wanted to do with it, he said: 'I wish it to be given to the poor of X' (the village in which they were working), 'especially to any poor families who have lost men in the present War.' The rest of the men followed his example, and between them they collected £2. 14*s*. 3*d*. There were under 200 prisoners, most of them being peasants from southern Italy.

You will therefore understand why I regretted seeing the following paragraphs in a Sunday newspaper, by a journalist[3] with a large public who in peace-time was an eloquent believer in the Christian virtues:

> Not long ago, somewhere in England, a deputation of women – typical nice country folk – came to see the commandant of a prisoners of war camp.
>
> This is the question they asked: 'What can we do to help your Italian prisoners, so far from their homes?'
>
> And this is the answer they received: 'Well, first you must ask yourselves: What can you do to help the soldiers of the Scottish regiment who are guarding the prisoners, as they are also far from their native heath and in a strange land?'
>
> I am afraid I should have said something much stronger myself. Wouldn't you?

Fortunately we are encouraged by the knowledge that all over the world there are merciful people who seek, however limited their power, to comfort prisoners and internees. British men and women, amongst others, are receiving this comfort. In a recent number of *The Friend*,[4] I read that an Army Captain on the staff of an Isle of Man internment camp had just heard from his sister, interned with her two children in south-west Germany, that German Friends had been very good to them. At Christmas they provided a Christmas tree, and presents for the children.

Often the refugees themselves, whether free or interned, are enrichers too. Like the Huguenots who came to England after the Revocation of the Edict of Nantes, they bring new industries and new ideas, and the younger ones give us back some of the youthful energy which the cities of Britain lost when thousands of children departed into the country or overseas. Even when many refugees were hurriedly interned during the summer of 1940, some of them went on regarding themselves as unofficial ambassadors to the country which treated them as 'suspects'.

One German girl from Glasgow, who had never known any home but Scotland since she was eight weeks old, was interned for eight and a half months in the Isle of Man because she was born in Hamburg. Her younger brothers and sisters, born in Britain, were left to carry on their normal lives. 'The memory of this experience will remain with me for ever,' she wrote to me, but added: 'In spite of all that has happened I bear no resentment whatever to those responsible for the too-hasty round-up – on the contrary, I am indebted to them for the wealth of knowledge I have gained from my experiences in an internment camp.'

Another internee, Dr Walter Zander, describing his nine months' internment as a Jewish enemy alien,[5] wrote that the most interesting aspect of this ordeal was not the suffering of the victims, but the extent to which they had been able to stand up spiritually to their trial and to transform their adversities into assets. He noticed the opening of this 'spiritual defence', on the morning after his internment, by a refugee publisher who sat down amid hundreds of tired, restless men to read the Bible for the first time in the original Hebrew. Another, a Greek scholar, began one sunny afternoon in the midst of tumult and anxiety to read aloud the song of Odysseus and Nausicaa from his pocket copy of Homer. Soon these men, and others of the same quality, were quietly carrying on their own shoulders the strain of the whole refugee community.

'It was somewhat like the fairy-tale', concluded Dr Zander, 'where a child falls into a deep well and finds at the bottom a wonderful green meadow; and the old truth became absolutely clear to me, that it depends largely upon ourselves whether or not we turn suffering into blessing.'

The danger, as he perceives, is that of leaving prison or internment as an angry, passionate rebel, anxious to 'get his own back' on the system which has caused his suffering. But in this case the humiliation will remain; honour will never wipe it out, for true honour and dignity are incompatible with bitterness, resentment, and the desire for vengeance. All captives whose conditions are not intolerable, whether prisoners of war, racial prisoners, or prisoners for conscience's sake, can help to solve the enormous problems of the present age by devoting their enforced leisure to the reading and thinking for which those who contend with war-time obligations have ever less time.

In these years during which the threat of prison has never seemed far away, I have often considered which books I would choose as permanent companions if I were arrested. If I were allowed half a dozen, I would select them, I think, in the following order: the Bible; Plato's *Apology of Socrates*; Dante's *Divine Comedy*; Bunyan's *Pilgrim's Progress*; Gibbon's *Decline and Fall of the Roman Empire*; and Tolstoy's *War and Peace*. Does it perhaps surprise you that I should put the Bible first? Some day you will find that its words, like those of all great literature, come suddenly alive under the impact of crisis. Verses which have become, through familiarity, a mere sequence of hackneyed phrases, leap from the page with a sudden relevance which illumines one's consciousness like a flame.

Take, for instance, the familiar passage from St Matthew xxiv, verses 7 to 14, which begins with the words: 'For nation shall rise against nation.' Could we, who are often isolated and despised, find a better description of our own anxieties than these verses, which had lost their sharp edge through repetition? Do we not all know friends who have been 'offended'; who have 'betrayed' us in ways which they rationalize as excusable; and whom we, in our exasperation and disappointment, have betrayed in turn? That 'the love of many shall wax cold', we realize regretfully of ourselves and others. The analysis goes home just because it illustrates those abiding conflicts which occur in every age, and have to be reconciled again and again. Surely, then, we can embrace with renewed courage the final promise of salvation for those who endure to the end? This experience of coming closer than ever before to the mind of God through the words of Christ makes the bitterest humiliation worth while.

But there are many too deeply impaired by the disasters of the War to be capable of accepting such consolation. They are the starving, dying millions of Occupied Europe, against whom, as Dr Nansen recorded after the last War, 'those who are ossifying behind their political platforms and who hold aloof from suffering humanity' have again steeled their hearts.[6] Their numbers are so great, their family lives so tragic, and their sorrows so fathomless, that we who have implored our Government to show pity on them by lifting the blockade sufficiently to permit the passage of some limited supplies, can picture their catastrophe only if we 'break it up small' and see it in terms of some beloved child on the verge of starvation. What should I do if for months I had watched you and S. getting thinner, paler, darker under the eyes, more quietly apathetic, until at last I knew that I had only enough for one more meal left in the cupboard? That is how I try to look at it.

We have to do what we can because no Nansen has yet arisen to save these starving peoples. In this War no Hoover Commission has been allowed to feed Belgium, where two million young Belgians are threatened with permanently warped minds and impaired physique. Only a few relief cargoes, sent from Turkey and Sweden, have so far reached Greece, where last winter 2,000 human beings died each day in Athens[7] and those who survived were so cold from hunger that they even opened the graves and robbed the dead of their garments.[8] These victims of the German occupation and the British blockade do not cease to die because the attention of newspaper readers is directed to Singapore or Cairo or Stalingrad.

It is difficult for this well-fed country to picture the results of starvation, and it must be impossible for productive and self-sufficient America. Yet the conscience of the United States has been more sensitive on this matter than our own. We in Britain, though kindly and well-disposed, are not an imaginative people. When we do get roused, we tend to be moved to righteous wrath by the crimes of other nations, while remaining indifferent to the victims of our own policy. Even now, few citizens of this Empire realize the basic similarity of Hitler's racial doctrine of a Nordic *Herrenvolk*, and our own complacent assumption of white superiority to the coloured races under our rule in Asia and Africa.

In the nineteenth century, our passionate sympathy with the Bulgarians and Armenians who suffered Turkish atrocities was combined with a singular insensitiveness towards the exploitation of the workers in British slums and factories, and with an almost universal ignorance of the oppression endured by native populations in British-controlled territories. Today our conscience is stirred by the crimes of the Nazis, but not by the European sufferers from our own blockade. We do not rid ourselves of responsibility by saying that their plight is due solely to Germany. In the case of Greece and Belgium it cannot be, since before the War both these countries imported over 40 per cent of their food.[9] If we could save the victims of famine, yet deliberately refrain from doing so, we are morally as responsible for their deaths as any Nazi conqueror.

Not long ago I was discussing our national 'blind spots' with a Quaker refugee from Central Europe, who after the last War helped to nurse some of the children suffering from tuberculosis caused by the food blockade of 1919. Other relief workers have described the small boys and girls whom they saw lying like limp rag dolls in the parks of Vienna because their limbs were too weak to carry them.[10] This woman told me that the flesh literally dropped from the bones of the children whom she nursed. She wrote to me afterwards:

> I cannot describe those sufferings to those who have not seen them. One boy, a child of nine or ten, had the face of a man of seventy, full of unspeakable suffering and patience; his arms and legs were only bones, partly covered with skin, but the larger part not at all covered, with the few muscles loosening from the bones, completely sore and skinless. Terribly painful to touch. His legs and arms were hung up by some contraption, because he could not have borne the pressure of them lying on the bed . . . This was one case. There were hundreds of them, in varying stages. Hundreds of boys and girls of all ages, two to sixteen, whose little bodies had deep holes, as if shot. This could be cured, but it would take six years or longer. I only describe this to you because you want to know the effects of starvation. It is *just hell.*

In 1922 this Friend paid a short visit to England. At that time anti-war reaction was growing, and many peace organizations were eagerly gathering members. One afternoon she joined a mass open-air peace demonstration in Hyde Park, and heard Maude Royden describe the effects of the blockade on Central Europe. As she listened to this account of the pitiful events which she herself had just experienced, my friend turned from the speaker to examine the audience. From the expressions – interested, indifferent, complacent – on the faces of those well-nourished men and women, she realized that not even the poorest member of the crowd understood the meaning of starvation. Her experiences had led her to dedicate herself to the cause of peace; she now perceived the difficulties of such work in Britain.

A recent incident in which I took part was not dissimilar. I had spoken at a meeting in Sheffield held to urge controlled food relief for Occupied Europe, when a questioner at the back of the hall made the point that the blockade and its effects were more merciful than the mammoth bombing raids which the R.A.F. had just carried out on Rostock and Cologne. He fully believed this, because he knew what bombing meant, but had never experienced starvation.

I agreed with him that the mass bombing of civilian populations is not only a questionable method of winning a war from the military standpoint, but leaves such grim memories of carnage that post-war relations are likely to be embittered so long as recollection continues. But bombing, at its worst, means the sudden death and injury of thousands of men, women, and children in a limited number of cities. Blockade involves the slow death by starvation of millions in many cities of many countries. Even the end of the War does not end its effects. For those who survive, it means stunted bodies, ruined health, and embittered characters; and its consequences may be passed on from generation to generation. That is why pacifists, few in number, universally dismissed – to use Nansen's words again – as 'fanatics, soft-heads, sentimental idealists', must do what they can to stir compassion. Too often, nowadays, compassion seems to be regarded as a form of Fifth Columnism. People fed on a diet of propaganda forget that it is pity which differentiates man from the sub-human animal who destroys his victim without feeling compunction.

I am no defeatist – that word which the critics of pacifism use so lightly, without any regard for its true significance. The only real defeatism is the belief – which we see all round us, but which I do not share – that human reason is unequal to the problems which confront it, and must therefore give way to force. I want to see our victory in this War. But victory for me does not mean acquiring the power to push other peoples into that outer darkness of desperation which gave birth to the ugly militarism in Germany and Japan that is now our Nemesis. It means the triumph of those spiritual qualities to which many people in this country seem as sadly indifferent as those whom we call our enemies – truth, justice, mercy, brotherhood. That is why I am ending this letter with a sentence which George Eliot wrote in *The Mill on the Floss*: 'More helpful than all wisdom is one draught of simple human pity that will not forsake us.'

Notes

1 The following significant words were used by Mrs Vijaya Lakshmi Pandit (Nehru's sister) in the course of a reply to an appeal made in 1941 by certain English women to the women of India:

> We are now told that it is the duty of our sons to go and shed their blood in a foreign land to defend the freedom which is in peril – and yet if they are bold enough to ask whose freedom it is they are to defend they are treated as traitors and spend long years languishing behind prison bars. Has freedom a double meaning – does it mean one thing for you and yours and something different for us?

Mrs Pandit was arrested and detained at Allahabad on August 13th, 1942.

2 November 22nd, 1941.

3 Godfrey Winn in the *Sunday Express*, December 7th, 1941.

4 February 13th, 1942.

5 *International Fellowship of Reconciliation Leaflets*, No. 7.

6 Dr Fridjhof Nansen's opinion of the politicians who had impeded his work for the victims of the last war for democracy was recorded in the speech which he made on receiving the Nobel Prize in 1923. The full quotation runs as follows:

> They represented that barren self-sufficiency, with its absence of any wish to understand other points of view, which is Europe's greatest danger. They called us fanatics, soft heads, sentimental idealists, because we have, it may be, a grain of faith that there is some good even in our enemies . . . I don't think we are really very dangerous. But the people who are ossifying behind their political platforms and who hold aloof from suffering humanity, from starving, dying millions – it is they who are helping to lay Europe waste.

7 *The Times*, January 22nd, 1942.

8 *Evening Standard*, January 28th, 1942.

9 For a full statement of the problem of food relief and the arguments on both sides, see Roy Walker, *Famine Over Europe* (Andrew Dakers). The author was recently sentenced to six months' imprisonment as a conscientious objector.

10 *Cf.* Maurice L. Rowntree, *Mankind Set Free* (Cape, 1939).

7

Youth and the War

There are always a few men and women who welcome a war. For some it is a source of financial profit or coveted political office; to others it offers a convenient licence for the exercise of cruel instincts. To still more it provides a way of escape from failure, boredom, unemployment, or insoluble domestic dilemmas. It is not of these that I am writing today, but of that great majority to whom war brings tragedy.

In my last letter I mentioned the prisoners, refugees, and starving multitudes of Occupied Europe whom pacifists have a special obligation to serve. I want now to write of three more groups – of which each member, don't forget, is an *individual*, with hopes, fears, affections, emotions, and aspirations exactly like yours and mine. They are the young men and women whose vital years of preparation for life are everywhere being stolen from them by the State; the children growing up, like you and S., in an age of great terrors, great problems, and a few great hopes; and the bereaved who mourn, as they mourned in their silent, neglected thousands 25 years ago, for husbands, sons, brothers, lovers, wives, daughters, and friends.

The ranks of the bereaved may be divided into two classes: those who are lamenting their dreams, and those who are mourning their dead. The members of the first category are mostly young. Both young and old belong to the second.

Among the innocent victims of this War, now amounting to millions, none is less guilty than the boys and girls born since 1918. Yet it is they who have to pay, in loss of life, health, hope, and opportunity, the heaviest cost of international breakdown. Like their parents between 1914 and 1918, they emerge from childhood to face exhortations to sacrifice themselves in the interests of a society created by their elders but not their betters. They are ordered to yield up their control of their own lives for reasons, however idealistically presented, which affect only the few who exercise authority. Even at 16, their registration makes them part of the potential 'man-power' or 'woman-power' at their government's disposal. Those stereotypes used so frequently by the Press – 'our troops', 'our airmen', 'reinforcements', 'prisoners', 'C.O.s', 'casualties', 'service women' – conceal a total sum of youthful suffering, bewilderment, and frustration never equalled in history.

Apart from the parents with serving sons and daughters, no one could be more conscious of the heavy heritage handed on to the younger generation than the fathers and mothers whose children have reached school age. When you, my son, were born in 1927, and S. in 1930 – that brief half-decade in which we seemed to be emerging from the aftermath of one great war without yet perceiving the shadow of the next – I still believed that I had saved both of you from the chaos, interruptions and losses of a world conflict by having endured those things myself ten years before. Instead, like all your contemporaries, you draw nearer each month to the choice between military service, with its sacrifice of education and career and its threat of death or damage, and a refusal which involves official dishonour, social unpopularity, and the same wasteful frustration. And you need to have evolved, prematurely, a complete philosophy of life before you are qualified to make that choice.

With reasonable luck, you may never have to make it. But others, only a few years your senior, are faced with it now. As the mother[1] of a grown-up family recently wrote:

> The young men who, through a quarter of a century, have been brought to perfection, young men, white, brown, and yellow, sink in the high seas and are lost, they perish in the desert and the snows, they fall from the skies and are destroyed, and their passing is lighted by the flare of burning homes. The day of total war is nearing its zenith.

Meanwhile a host of young women, who should have married them and borne their children, is again condemned to sterility. Today, too, the young women are conscripted themselves. They are urged to put the War machine before the 'sacredness' of family life, though this 'sacredness' was always the reason given in peace-time for withholding from married women the right to continue working in the professions for which they had been trained. No one troubles about two incomes, or five, going into one home when that income is made through war service. Only when it is earned by such creative occupations as teaching and medicine are objections raised. The destructiveness of modern war strikes more fiercely than ever before at the things that mean most to women – children, homes, education, healing.

Even from the young men, neither death nor injury is the only contribution which the war-makers demand. The prospect of prolonged, indefinite tedium saps the mind and spirit as surely as starvation exhausts the body. Today's youth is wasting its best years in suspense and monotony. It waits about. It goes on duty to do nothing of importance for leaden-footed hours. 'Not even an Alert to liven us up!' a 22-year-old girl Civil Defence worker wrote to me in exasperation after months of inactivity. Sometimes the experience of disintegrating uncertainty starts even before the period of service begins. The attempt, for instance, to combine military training with a university education has not been precisely helpful to either. 'Why is it', inquired a contributor to the *Daily Mail*,[2] 'that thousands of young men and young women have to be made dispirited and disgruntled, waiting for months after they have been told that they may expect their call-up at any moment?'

Yet, comprehensible as these reasons for frustration are, I believe that its real explanation lies deeper in the psychology of youth. I should define it as a vague, uninformed, yet growing scepticism regarding the justification for the War, and an indeterminate yet pervasive doubt whether its alleged objective is really worth the loss of so much that makes life desirable. Even to the unreflective, the merely negative 'destruction of Hitlerism' makes little appeal, while 'survival' offers still less. Among the more intelligent I find a profound misgiving whether, even assuming that the Nazi forces could be utterly annihilated, this in itself would 'destroy Hitlerism'. The fact is slowly being realized that 'Hitlerism' is something far deeper, more intractable, and of longer standing than Hitler himself. Many young people now recognize that the Nazis, like the Bolsheviks, are not the cause of the chaos to which our civilization has come, but part of its effects; and that the remedy lies, not in further threats of punishment and repression, but in the promise of a future which holds out some hope to the weary peoples of Central Europe. The young and generous are ready to make such a promise. But their leaders fail to perceive the necessity.

Neither the young men nor the young women do much complaining. Only a few resist the ageing politicians who send them out to fight rather than agree to the sacrifice of material benefits (which is the basis of all negotiation). Often, in spite of the death-dealing weapons placed in their hands, they show, like companions in misfortune, spontaneous humanity to their 'opposite numbers'. In

January 1942[3] I read of an Italian doctor at Halfaya who brought five wounded bomber pilots back to the British lines because he had no medical supplies with which to treat them. In return he was given an abundance of surgical dressings to take to his own men.[4]

One afternoon in the first autumn of the War, I learnt that the young of today do not die joyfully for the mistakes made by their predecessors. I had gone to Eton to have tea with a master whom I knew at college, and we were sitting in his room when three boys in khaki came in to say goodbye. There was no jubilation, as there would have been in 1914; not a flicker of excitement nor hint of congratulation. Master and boys alike were inarticulate and sad. Incongruous as the background to that sombre acceptance, the scarlet ampelopsis climbing the grey walls and the river sparkling in the autumn sunshine took on an indescribable poignancy. I did not learn where the boys were going, but after they had left the master told me that they were his star pupils of the previous summer. Whenever I recall them now, I remember a verse from Edna St Vincent Millay's *Dirge without Music*:

> Down, down, down into the darkness of the grave,
> Gently they go, the beautiful, the tender, the kind.
> Quietly they go, the intelligent, the witty, the brave.
> I know. But I do not approve. And I am not resigned.

No doubt you will say that Eton and its inhabitants are typical only of the few. But I have found little evidence of greater enthusiasm in the East End of London. A Bethnal Green settlement worker once told me that in this War no glamorous aureole crowns the young soldiers who come home on leave; their juniors pity rather than envy them. Now and then Etonians and East-Enders alike break forth privately into passionate words. I had one such letter from a 23-year-old lieutenant a few months ago:

> I want to try to explain the feelings of some of those taking part in this War. The youth of this age began this War mentally at the stage where those engaged in the last War left off; no surges of heroism, we had seen too many without limbs or permanently mad from head wounds, no belief in good versus evil, it is grey versus black, or tolerable against the intolerable, and very little hope of there being a bigger, better, and happier world after this War has ended. I know that all this is common knowledge to everyone except those who make speeches . . . The fever-pitch of enthusiasm that 1914 evoked made facing death far more easy, the knowledge of the inevitable

> incompetence that causes most of the casualties was not present to shake the trust in one's leaders . . . The general atmosphere has so radically altered, the amount of times one hears that it really does not matter whether I come back alive or not, for things will inevitably be so hellish after the War that it is just as well that I should cease to be, is frightening . . . At Oxford I had a friend who was extremely intelligent, good-looking, and popular, besides being senior scholar of his year at his college. He was old for his age, and thought; he got so hopeless that he took to drink seriously (one bottle of whisky a day) simply because he was unable to solve the uselessness of life; weak, you may say, but he was not alone. Now there is a war on he is flying and can forget his sorrows in another way, but as to militant nationalism it does not enter into things . . . I am certain that the generation that *experienced* the last War must get in touch with those who have been in this if any good is to be achieved in the future.

A few weeks later, a letter from another stranger, this time in Egypt, confirmed the observations of the first.

> It seems very hard [wrote this young soldier] that there was not a strong enough body either in England or France to demand an expulsion of the reactionaries and the 'old men', and so save a second generation from this sinister doom. The worst part of this War seems to me the utter lack of hope for the future. If I could feel stimulated – as I understand so many young men and women were in 1914, believing it was a war to end wars and that we were like chivalrous knights of old, ready to give our lives so that others might live in peace and happiness – I feel I might be able to work myself up to a fervour of self-righteous elation which would at least be preferable to this deadening sensation of 'just grinning and bearing it' in order to avoid the fate of Poland and all those other victims. Do you see any guarantee that my generation is not doomed to the same fate as yours, and is to be thrown to the wolves with no further prospects for the world than mass starvation, criminal statesmanship, and more wars?

These letters and many others I could quote make me certain that the rescue of our society, and especially its youth, from the sense of frustration and disillusion now possessing it will never be achieved by further exhortations to damage and destroy – however heroic war propaganda may cause these performances to appear. Young adult human nature is essentially constructive. One reason for the present despondency of boys and girls in the Forces, and of those not yet mature enough to join them, is the failure of the Government to decide what type of society it wants to see after the War, and its reluctance to guarantee the changes which are necessary if the new is to be better than the old.

It is a common practice now to vilify France and denounce her collaborationists, but, according to a contributor in a recent issue of *Blackfriars*,[5] she has at least created 'a great body of young people whose one aim is service, whose watchword is generosity, and who are ready to make all the sacrifices necessary to help their suffering fellow countrymen'. Today these boys and girls who represent the real France are directing traffic, housing and feeding refugees, giving information, carrying luggage, delivering the mail in the place of a defunct bureaucracy. They are the products of a Catholic revival which has helped the Vichy Government to function by creating this fine type of youth at the moment of France's greatest need. Their part in rebuilding France has saved them from the fate of many unhappy young refugees whom total war has cast upon the world. Other children have lost all who were left to care for them before being themselves deprived of a life which began only to end. A few weeks ago I came across this short paragraph,[6] which was headed 'The Unknown Child':

> A baby girl found unconscious in air-raid ruins at Bath died a few hours later. No one knows who she was; no one has claimed her. The Rev. H. K. Hudson, formerly Vicar of Berden, Bishop's Stortford, has suggested that the child should be buried in some ruined church as 'the unknown child victim'. Every possible effort has been made by the authorities to establish the baby's identity, but they have failed. Her coffin bears just a number.

For many young evacuees, the choice has lain between nights of terror and days without kindness. Only a fortunate few, like you and S., have had generous American friends who offered a sanctuary rich in affection. During the years between the wars some of my correspondents, writing to me of their war-time childhood, mentioned dark recollections of tension, of telegrams, of blackened windows, and decreasing rations. But what will the memories of today's children be – many hounded from country to country, still more wakened by guns and bombs, others starved to the point of death, yet others forcibly severed from beloved families and cherished homes? How can we, your parents, who want to see you confronted, before we depart, with a future at least better than our past, turn these recollections to account? What can we do to ensure that out of them shall spring, not cynicism nor vindictiveness nor pessimism, but the constructive determination to save yet another young generation from your own harsh experience?

However hard your future may be, I believe that it is only by carrying on the struggle to create a lasting peace that you and your contemporaries, both older and younger, will find compensation for your insecurity, your disillusion, and your memories. Whether your contribution is made through art, through letters, through welfare-politics, or through day-by-day experiments in creative living, matters very little. What does matter is that we, your predecessors – parents, foster-parents, preachers, and teachers – shall make as certain as we can, by the education we give you, that you do grow up to fight against warfare, and not, like yesterday's children, against one another.

Notes

1 Barbara Duncan Harris, in the *Monthly News Sheet* of the Women's International League for January 1942.
2 F. C. Hooper, *Daily Mail*, February 12th, 1942.
3 *Daily Mail*, January 8th, 1942.
4 In August 1942, a letter received from a stranger, a private serving in the Middle East, referred discreetly but unmistakably to a similar experience with German troops:

> Some time ago I wrote to a friend at home about hatred . . . I said I would never agree with the Vansittarts and other elder statesmen about the character of Germany as a whole . . . Not by gathering my own observations, but by hearing the opinions of those who shared this experience with me, has the above been proved right. If collaboration occurs during battle, how easy will collaboration be after war. If all our forces had had similar experiences, I doubt whether the War would have continued. It proved to me also what I have always thought is essential for peace – a vast interchange of youth, for let youth meet and all hatred and misunderstanding will disappear . . . I have only one wish, to tell Vansittart & Co. how utterly wrong they are, for I know none of them could ever produce such overwhelming evidence.

5 Blackwell, Oxford; quoted in *The Christian News Letter* for January 21st, 1942.
6 *Evening Standard*, June 19th, 1942.

8
They that Mourn

The mourners of today, though they form a worldwide multitude, are not easily discovered by their would-be comforters. As I wrote you in my second letter, one purpose of modern war propaganda is to conceal personal suffering as effectively as it hides the horror of every war-time 'incident'. To the popular collective nouns in which individual losses are submerged, the Press and the *communiqués* have added a now familiar series of conventional phrases designed to reassure by their studied vagueness.

'Mopping-up operations are proceeding', we are told, 'and our troops are cleaning up pockets of enemy resistance.' 'Some damage was done, but casualties were negligible.' 'The situation was always under control, and all services worked smoothly.'

I once read an article[1] analysing the facts so comfortably disguised by the first of these amorphous expressions. ' "Cleaning up pockets of enemy resistance" reminds us of the swift, health-giving action of the surgeon's knife,' comments the writer, who by now probably knows the exact stage of an argument at which his companion is going to describe this war as 'a surgical operation'. 'It does not suggest a pitiful group of Italian peasants, exhausted by fever and dysentery, whose only existence is the short, blissful silence between machine-gun and bombing attacks.'

In the East End of London, and other raided areas which I knew during the Blitz, I learned and not infrequently observed what happened to the 'negligible casualties' in those well-controlled 'situations', and saw the distress of their husbands, wives, fathers, mothers, or children. Let us hope that there are many future writers and artists among the shelterers, wardens, fire-fighters, and ambulance-drivers, who will one day make real the meaning of this War in human terms. They alone have the power to show us the lost homes, the broken families, the sorrowful wives, the terrified children, who are pawns in the huge political chess-game of total war. At present the griefs of these families find expression only in the correspondence of their scattered members.

> My wife and two sons are in Ottawa [an old acquaintance wrote me the other day from the Middle East]. I, after exploring every possible avenue for a decently paid job, found myself forced to apply for a commission, for the second time adopting a career to which I am about as ill-fitted as possible. Army life is definitely unsuited to a bald bespectacled suburban family man of 47! . . . Sweat, flies, sand, boredom, make up the days and nights. Meanwhile one's children are passing through the most interesting stages of their development and one is missing it all. Sometimes I feel as if I should never get over this split up of my family.

Once, even in the last War, sorrow was honoured as an emotion deserving of consideration. This respect was extended to the mourners of 'the other side'. Today even that measure of human decency has vanished. Few forms of propaganda have supplied such disquieting evidence of lost standards as our imitation of the savage Russian broadcasts designed to turn the screw of suffering in the hearts of bereaved German women. One leading article in an evening newspaper was entitled 'Mrs Hess'. It was addressed in taunting terms to the woman whose torturing suspense at the time of her husband's strange flight is shared by every wife of a British commando.

If these cruelties are ingredients of morale, we are better without it; it becomes too crude a substitute for our former ethical values. We can forgive the Russia of the Revolution, with its totalitarian order built by blood and terror upon a foundation of semi-Asiatic barbarism. But ours has been nominally a Christian civilization for centuries. Though we have never actually practised Christianity, we have officially accepted the standards of a faith which enjoins us to forgive our enemies not seven times, but seventy times seven.

Here, as in Germany, they that mourn must today find their own consolation. They will discover it, perhaps, when they are permitted to speak again, in making their pain real to our defective imaginations. Some never find it at all, and take the only means of escape that remains. Not long ago I saw this item, typical of many others, in the evening Press.[2]

> A few days after hearing that her R.A.F. son had been killed, Mrs Irene Bratley, 50, was found dead in a chair in front of an electric fire at a flat at Marine Parade, Brighton. At the Brighton inquest today, when a verdict of suicide while the balance of the mind was disturbed was recorded, it was stated that Mrs Bratley was discovered by police

> on Monday after the glow of an electric fire had been seen from a bedroom on successive nights. Dr L. R. Janes, pathologist, said that death was due to poisoning by a barbituric acid derivative. The Coroner, Mr Charles Webb, said that a note left behind contained the sentence: 'Please bury me as cheaply as possible, and don't waste any tears. I am longing to go.'

The majority, of course, stop short of suicide. They go on living from force of habit, rather than because they see any reason for hope. Recently I received a letter from an unknown reader, this time a girl in her twenties. She wrote to ask for help in publishing the manuscripts of 'a very beloved elder brother'. He had been a pilot officer who disappeared, with all his crew, on his first flight against German shipping. Heartbroken, she turned for comfort to the man she hoped to marry, but overwork and war-strain had warped his mind, and they parted. In desperation she transferred all her hopes to her youngest brother, a gay, handsome boy of 19. 'He understood, and did his utmost to fill the gaps. He went down with his ship in January and my life has just been wiped out, the three I cared for most in the world gone, and the future just a blank.'

Is it surprising that the Cardinal Archbishop of Ireland made a plea to the statesmen of the world for 'the poor plain people . . . to whom victory on either side will not mean very much'? Or that Dr Alfred Salter, M.P., quoting him in the House of Commons, concluded his own speech with another appeal: 'Is there to be no end to this torture of millions of human beings? Is there no pity in the whole world?'

More recently, broadcasting on the twenty-fifth anniversary of his ordination as a Bishop, the Pope issued another grave reminder to the organizers of continued war. Behind the war-front, he said, arose another huge front, the front of family anguish:

> We should like to address a fatherly warning to the rulers of nations. The family is sacred. It is the cradle of children, and also of the nation, of its force and its glory. Do not let the family be alienated or diverted from the high purpose assigned to it by God . . . We think of the separation between husbands and wives, and of the destruction of family life; of famine and economic misery. There are heart-breaking and unending examples of every one of these. This is one of the most terrific and terrible things which has ever happened to mankind; indeed, it is such as to make us fear grave economic and social dangers for the future. While great intelligences were busy trying to build a new social order, and people knew that national

> wealth was one of the fundamental bases of the solution, today this national wealth is being spent by the hundred millions to destroy all that exists.[3]

We, who cannot join in this work of destruction, must learn, as part of our obligation to a suffering community, what to say to those that mourn. 'Passing through the Valley of Weeping', wrote the psalmist, 'they make it a place of springs.' But perhaps you think that just saying something is not much good? In these noisy, over-exciting days, people are apt to speak of words with contempt. They must have deeds, they say. Nothing matters but 'constructive action'.

I do not agree with them. The power of words is greater than anyone can calculate. Words have made revolutions before today. The use of the right words at the right time can transform the existence of a man or woman from desolation to glory. But you may have to live a whole lifetime before you learn how to choose those words.

It therefore remains true that, for many young and active pacifists, the best outlet for spiritual tension is practical service to the victims of power. Often the tension is released most effectively by work which no one else will undertake because it is so unpleasant or so unpopular. During the air raids many members of Pacifist Service Units did the humblest of necessary jobs in the shelters, washing floors, disinfecting bed-clothes, or collecting the dirty scraps left by the shelterers. One unit, organized by the Anglican Pacifist Fellowship, decided to open, in a deep basement near Charing Cross station, a refuge for the 'down-and-outs' who normally sleep in the parks or on the Embankment, and were not accepted in the ordinary shelters.

These social outcasts had already taken informal possession of this dark sanctuary when they were discovered by some young pacifists working in the crypt beneath St Martin's-in-the-Fields. The friend who showed me round the now civilized refuge known as the Hungerford Club described its original appearance: he spoke of the fire-buckets lit during the cold winter of the Blitz in the gloom of the long covered corridor; the dirty hands stretched over the glow; the dark, suspicious faces redly illumined. That flash of sinister drama has vanished; order, gaiety, decency have taken its place.

Because the members of this Pacifist Service Unit have known humiliation, suffered as social pariahs, and, in one or two cases, been to prison, their skill in establishing friendly relations with London's vagrants has been quite unusual. The inspirational genius which enabled Canon Sheppard to make the crypt of St Martin's a refuge for homeless wanderers is at work in the Hungerford Club, but it has a different origin. His natural gift for human understanding has here been created by stark experience of shared degradation.

Between these once dismal but now cheerfully painted walls, 'humiliation with honour' is demonstrated in action. No normal social workers, gifted perhaps with natural kindness but endowed with a subconscious sense of elevation through lack of reciprocal knowledge, could have made the same success of their job as this group of conscientious objectors. Perpetually handicapped by lack of funds, often infested with vermin passed on by their visitors, sometimes bruised and battered by tussles with 'drunks' who subsequently return to apologize, these workers close to Charing Cross have maintained an unimpaired friendship with their guests and with each other. The drunkards have not ceased drinking nor have the drug addicts escaped from their bondage, but the 'regulars' who were once dangerously filthy are now comparatively clean, and the young men who sleep at the Club are as safe from attack as anywhere in London.

> We cannot shelter behind the decision society has passed upon them [runs the latest report of the Hungerford Club on its 'undesirables'] unless we are confident that the verdict of man is upheld in the courts of heaven. Therefore, although these folk may be little credit to the human family, they still belong to the family of God and, in the name of the Great Brotherhood, can press their claims upon all others who own the Common Father.

During the summer of 1942, a young woman pacifist found a way of bringing comfort to another and more temporary class of 'undesirables' – the friends and relatives of the Nazi raiders who have died in this country. Passing through the cemetery of a recently blitzed town, she came across the graves of three German airmen brought down in the attack.

> They were overgrown with weeds and completely neglected [she wrote to me]. Remembering the care and devotion with which some friends of mine tend the grave of their own airman son, I couldn't help thinking of those other parents, or perhaps wives, over in Germany, who couldn't pay even that last tribute to those whom they loved, and who, however mistakenly, had given their lives for their country.

She resolved that she would tend the graves herself, and asked if I knew of any way by which this information could be conveyed to the relatives of the airmen. An attempt to communicate through the Red Cross proved unavailing, but thanks to the initiative of my correspondent the attention of the responsible organization was drawn to these graves, which are now beautifully tended. Sooner or later this news will reach those German families who are mourning the death of their sons in an enemy country. Thus, through the imaginative sympathy of one young woman, another of those personal gestures has been made which create a worldwide fellowship amongst those who suffer, and link the 'poor plain people' to one another across the chasm of war.

Notes

1. *Peace News*, January 2nd, 1942.
2. *Evening Standard*, February 13th, 1942.
3. *The Tablet*, May 23rd, 1942.

9
The Descent of Society

In my fifth letter I suggested that an important function of pacifism in war-time is the preservation of certain human and religious values which might otherwise be lost. These values are vital to the making of peace, whatever the circumstances in which the struggle ends. Their abandonment at the end of the last War meant the failure of the Treaty, in spite of an overwhelming Allied victory.

I also emphasized that this undertaking must be carried out humbly and without self-righteousness, since we are all part of a society which has failed, and must bear our proportion of responsibility for that failure. Their forgetfulness of their own share in the present disaster makes some pacifists assume a superior air which exasperates those who do not agree with them. To follow a light which you have seen may compel you to a certain course of action, but this does not in itself make you a better human being. The private lives of pacifists are often difficult just because their public witness makes perpetual nervous and emotional demands. Even the beloved leader, H. R. L. Sheppard, as his biography by R. Ellis Roberts reveals, could not save his household from paying indirectly a heavy part of the price of his faith.

The price must nevertheless be paid. Even under favourable conditions, pacifism is always a discipline of the soul. In time of war it becomes a very stringent discipline, for the majority of its critics deny the spiritual compulsion from which it springs, and attribute it to every type of unworthy motive that ingenuity can invent.

To withstand these critics and maintain a detached judgement in spite of the sense of injustice that they cause and the popular support that they command, is a moral exercise which greatly strengthens the fibre of those who can endure it. Criticism makes it only the more necessary to keep our heads and continually review the events of the War in the largest possible perspective against their historical background. This perspective becomes increasingly obscured by war propaganda, yet only by seeing it steadily through the fog can we serve the cause of truth. Facing facts as they are to the best of our ability, we must convey the knowledge of them to others by every means at our disposal. Through the shadows of these days we struggle towards the light cast in dark places by love, knowledge, and pity, and so fight to the limits of our power for the better standards against the worse.

My last three letters have shown you how unlimited is the human misery which stakes its claim upon compassion today. I have described how a war begun – at any rate by some – in the genuine belief that it would save the persecuted groups in Central Europe from Nazi tyranny, has resulted in carrying suffering to the people of almost every country on a scale which even the worst negotiations and the least favourable peace would never have approached.[1]

But suffering, as these letters have tried to explain, may be redemptive if it is accepted without bitterness as a part of experience. Pain, loss, grief, failure, degradation, and oppression can and often do carry an individual to spiritual adulthood more swiftly than the most gratifying encounter with joy. The same cannot be said of the deterioration of moral standards which accompanies war. For this there is no compensation; it produces worse men and women and a more evil society. Much of the work of pacifism in war-time involves exposing and fighting this process, and thus trying to prevent the spread of spiritual rot.

The values which I wrote you that we must endeavour to preserve mean an unswerving allegiance to Christian standards of charity, magnanimity, compassion, and truth. These weapons of the spirit offer a constant challenge to the easy war-time descent into animosities which damage the hater far more than the hated. They combat the vengeance which disguises itself as 'retribution'. They resist the ruthlessness which inflicts suffering without compunction upon the helpless and weak. They unmask the propaganda which denies, distorts, and suppresses facts. There is no end to the moral damage wrought by war-time passions. Many waters, it is said, cannot quench love, but it perishes quickly in the miasma of suspicion and fear. All sense of human claims vanishes, as my second letter showed you, in a pitiless, impersonal collectivity. Soon the very power to feel becomes itself impaired. The death of men's sensibilities is one of the earliest and most disastrous casualties in war.

Between 1914 and 1918 I learned from my work in military hospitals the numbing effect of perpetual contact with suffering. During the opening days I felt sick at the sight of a raw wound, and the first death that I witnessed haunted my mind for weeks. Before a year had passed the wounds to be dressed represented merely so many hours of routine, and I would calculate whether a mortally wounded patient would die in time to give me my weekly half-day off duty. Yet I am not, I think, brutal or unimaginative. The continued experience of tragedy gradually atrophies the senses, and creates a defence mechanism of callousness without which the normally sensitive person could not endure.

With the less than normally sensitive, this process soon degenerates into barbarity. With all of us, actions once unimaginable come to be accepted as normal, or, at worst, unavoidable. 'So Lübeck was bombed', commented an editorial in the *Daily Herald*,[2] formerly a newspaper which defended the world's workers against the will of the privileged few. 'In the course of the bombing humble homes were devastated, innocent civilians killed. *But that cannot be helped*.'

At a recent Conference, Lord Ponsonby found the comprehensive word which describes the external symptom of our deadly disease: 'Acquiescence: that is the evil.' And acquiescence takes a long step down a slippery slope, for the next stage after the loss of compassion is the infliction of cruelty – always justified on the ground that the enemy is worse. It can hardly be over-emphasized that the endurance of pain, however bitter and unprovoked, never rots our moral fibre with one tenth of the speed that follows its infliction. Finally comes the deliberate cultivation, by hitherto civilized persons, of the viler qualities which are latent in us all, and the considered disparagement of their opposites. The finest human attributes are dismissed as valueless in terms of contempt, which classify magnanimity as 'effete' and describe a belief in the power of forgiveness as 'moral aerobatics'.

In 1939 the democracies opposing Hitler accepted as their standard of war-time conduct the regulations of international law, which is based upon the belief – fully justified by the familiar psychological process described in the previous paragraph – that it is better to suffer disadvantage in war than to descend to the lower levels of barbarity. By common consent such barbarity included the slaughter of civilians by bombing or starvation, and the avoidable destruction of humanity's cultural heritage. In 1940 a group of pacifist clergy went on a deputation to the Archbishops of Canterbury and York. Part of their purpose was to inquire at what point the Church would rather see the War lost than won by methods inconsistent with Christianity. The Archbishops replied that this point would be reached if the bombing of open towns were undertaken, not as a reprisal, but as part of our national policy; and also if we deliberately violated the neutrality of another nation.[3]

After three years of war, the comparatively civilized position adopted by both Church and State has degenerated into one in which government spokesmen brought up as Christians describe the sending of mercy ships through the blockade to feed the starving children of Europe, as 'false humanity'; and once kindly listeners to the radio revel in the massacre of civilians by the thousand in Rostock or Cologne. The 'plastering' of a selected area has replaced the 'precision bombing' of military objectives, and Air Chief Marshal Sir Charles Portal addresses an extraordinary admonition to the readers of Sunday newspapers: 'Let it not be imagined that the effects of this bombing are confined to the civilian population.'[4] British troops have occupied Iceland, Iran, and Madagascar. Yet the point at which the Church was to have found the War intolerable for Christians has not yet arrived.

In a letter to the Press during 1941, Professor Gilbert Murray and Mr George Bernard Shaw pointed out that reprisal bombing is not only morally indefensible but militarily useless, since even if raids could be maintained nightly, and each raid killed 1,000 persons, it would take over a century to exterminate us, and a century and a half to exterminate the Germans. The Military Correspondent of *The Times* has calculated that in order to bomb Germany as intensively as undefeated Malta was bombed during three weeks of 1942, we should require over four million bombers.[5] The Church does not support these critics even for reasons of common sense. Lesser mortals protest against the mass bombing of German cities on the grounds that we are now guilty of the very type of conduct for which our Government went to war against Germany. An editorial in the *Daily Herald* describes these objectors as 'woolly-headed' – although, if they deserve this term, it is equally applicable to Hugo Grotius and the other founders of international law. But the Church is still silent.

'We have no apologies to make for devastating Lübeck,' the B.B.C. assures the Continent after the destruction of an old and lovely city which contained many irreplaceable treasures of medieval culture.[6] 'Blast and bomb, attack and attack', echoes the Press,[7] 'until there is nothing left where once men lived and worked. Do the job thoroughly. No sentimentality, no half-measures.' Even *The Times* reports with apparent satisfaction: 'It is estimated that already not fewer than 1,000,000 persons throughout the Reich are homeless in consequence of R.A.F. raids . . . The official German news agency says that Mainz Cathedral was among the historical monuments burnt down after the two successive air raids on the town.'[8]

On July 28th, 1942, Air Marshal Harris, the Chief of Bomber Command, broadcast to Germany a 'message' singularly unBritish in its boastfulness:

> We are going to scourge the Third Reich from end to end . . . We are bombing Germany, city by city, and ever more terribly in order to make it impossible for you to go on with the War. This is our object. We shall pursue it remorselessly . . . Already 1,000 bombers go to one town, like Cologne, and destroy a third of it in an hour's bombing . . . No part of the Reich is safe. In Cologne, on the Ruhr, at Rostock, Lübeck, or Emden, you may think that already our bombing amounts to something. But we do not think so. As our own production of bombers comes to a flood, and as American production doubles and then redoubles, all that has happened to you so far will seem very little . . .

The following Sunday, a commentary by John Gordon, Editor of the *Sunday Express*, endorsed both these words and these methods. 'Germany,' he wrote, 'the originator of war by air terror, is now finding that terror recoiling upon herself with an intensity that even Hitler in his most sadistic dreams never thought possible.' The longer you think over this sentence, the more astonishing its implications become. The most relevant comment is a question: 'Who is the sadist now?'

Would any impartial neutral regard these utterances as differing in any way from the hysterical threats of Dr Goebbels? Can any rational person consider the hopeless and terrible prospect envisaged by Air Marshal Harris and Mr Gordon as offering to the German people an incentive to abandon their only leaders, however deeply they may dislike them? But the Church, like the B.B.C., has no apologies to make. The Christian 'sentimentality' which it has taught for centuries is not applicable to total war. It is content to let its standards slide downhill with those of the Government and people over whom it claims the moral leadership.

Hatred is now much more than the moral laxity which accompanies warfare; it has become a policy. This stage began when the authors of the Atlantic Charter, visualizing an agreeable Never-Never Land in which the wishes of British and American statesmen would be carried out against the perfect background of a political vacuum, proposed in Point 8 to cure 'aggression' by allocating another dose of humiliation to the nation whose reaction against the humiliations of the last War produced the present catastrophe. By the official adoption of 'retribution' as a major war aim, this policy has been continued. It has reached its peak in the campaign of deliberate hatred which seems likely to confer an inglorious immortality upon the name of Lord Vansittart. This campaign is itself an example of inverted Nazism, a form of moral intemperance which is better guaranteed than any movement now in progress to keep Germany implacably united behind Hitler. Such policies would never have attained their present dimensions had the distinction, so carefully observed by editors and politicians in the early days of the War, between the Nazi leaders and the German people been resolutely maintained. Stalin himself has usually been careful to avoid the grave psychological error which identifies the two.

The Vansittart movement ignores facts as crudely as it abjures benevolence. Lord Vansittart's pamphlet, *Black Record*, sacrifices objective historical accuracy with a consistency worthy of a nobler purpose than the desire to inflict retaliation upon a people amongst whom the author was unhappy in his schooldays. The size of his following would have been impossible three years ago. It exists today because, in the realm of truth as in the field of charity, the British nation has passed from respect through indifference to repudiation.

In war the denial of truth begins earlier than the betrayal of pity, for truth is less perceptible than charity and its abuse is therefore harder to detect. From minor omissions and misrepresentations, Press and radio pass to major suppressions and exaggerations, and thence to false news and deliberate inventions, such as the 'Corpse Factory' story of the last War.[9] In this case the Chief of the British Army Intelligence is said to have interchanged the titles of two captured photographs, one picturing German soldiers being taken away for burial, and the other showing dead horses en route to the soap-factory. Evidence already exists that a similar manœuvring with captions is being practised to-day.

In April 1942 the Ministry of Information issued a picture showing a Russian supporting the limp body of a woman with her arms outstretched. At their feet lies the corpse of a young man. The original title under this picture ran as follows: 'German atrocities in Russia. Parents find the body of their murdered son in Kersh.' But when it appeared in the *Sunday Pictorial* it was 'improved' by some colourful embroidery:

> 'It's My Son . . .' This is a poignant picture. The Germans, driven from Kersh, murdered many of the Russian inhabitants in cold blood before they left. Those Russians who escaped from Kersh return to find their relatives. This mother has searched among the bodies of the slain, dreading to see a well-loved face upturned in death. Her worst fears are realized. She finds her son slain and, in an abandonment of grief, flings wide her arms and cries her anguish aloud.

Another newspaper, *The People*, in publishing this picture, departed from the Ministry of Information 'line', embroidered or plain, for a thesis of its own: 'The Tragedy that is Russia. A Russian father raises the murdered body of his daughter in Kersh.'

A final comment on the picture's adventures appeared in *Truth*:[10]

> It is this kind of reckless propaganda which makes popular journalism stink in the nostrils of honest men and lays British honour at the mercy of Herr Goebbels. There is surely enough evidence of the horrors of war as Germany and Japan conduct it without these imaginative excursions into forced melodrama. For all we know, the picture may actually be that of an episode in the Russo-Finnish or any other campaign.

This picture is only one example of the war-time abuses of truth. There are countless others, such as the suppression of our own reverses and the proclamation of the enemy's; the presentation of his conduct as incessantly devilish, and our own as perpetually noble; the 'spotlighting' of one incident in a *communiqué* so that the whole picture is out of perspective; the selection, on Vansittart lines, of certain aspects of a nation's history while omitting the rest. It is the peculiarities of human psychology which make atrocity stories so potent and so dishonest a weapon. Most people tend to be supersensitive to certain types of 'atrocity', and to remain singularly impervious to others. This enables the Press and the B.B.C. to present our own brutal deeds with triumphant complacency, while displaying fervent moral indignation over the cruelties of the enemy and carefully omitting any information which might mitigate or explain them.

For weeks during the spring of 1942 we were urged to 'remember Hong Kong', for if atrocities did not exist in war, it would be necessary for governments to invent them.[11] Shortly afterwards, an acquaintance of mine arrived home from Singapore. She had lost her home and had left her husband in Japanese hands, but her mind remained detached and unembittered. The Japanese, she told me, had suffered severely at the hands of our Gurkha regiments, whose cruel instincts the British had not restrained. This fact, if true, does not undo or excuse the Japanese brutalities, but it places them in quite a different perspective.

There is, of course, no reason – apart from the liability of human beings to divide their minds into watertight compartments – why the Japanese excesses in the Far East should be labelled 'atrocities', while the bombing of civilians, and the starvation (whether by pillage or by callous acquiescence) of children in Greece and Belgium, should masquerade under more dignified names. Any act of cruelty against the helpless and defenceless is properly called an atrocity.

The persons most readily moved to vindictiveness by excited stories of brutality are usually non-combatants, particularly the safest, the most comfortable, and the least occupied. During the spring of 1941, when a Gallup Poll was taken in England on the subject of reprisal bombing, the largest demand for reprisals (76 per cent of the population) came from Cumberland, Westmorland, and the North Riding. In Inner London, which had suffered severely, the supporters of reprisals were only 45 per cent.[12] Their bitterest advocate, in my own experience, was a middle-aged woman living securely in a private hotel at Shrewsbury. Those who fight and those who suffer have neither time nor energy for hatred. They experience in their own flesh and nerves the pain of the enemy, while every new attempt to damage him recoils upon themselves. The rest make the psychological mistake of seeking to destroy the external 'foe' who disturbs their peace, instead of combating within their own souls those evil emotions which lead to more and more wars and the moral decline that follows them.

How, you may ask, can a small, unpopular, and restricted minority do anything to arrest the spiritual descent of a nation? Its voice is weak; its deeds are condemned or disregarded; its limited writings are criticized and controlled. When there is a shortage of the very commodity from which books, the vehicles of opinion, are produced, it is naturally allotted to those subjects which command the maximum support.

Our powers, it is true, are circumscribed, but the Scriptures themselves have urged us not to despise 'the day of small things'. The duty of protest alone requires constant alertness. But protest, though it may demand more courage than any other kind of witness, remains essentially negative. Charity, not protest, is the chief weapon of good against evil, of peace against war. But charity today is officially regarded as one of those inconvenient virtues which must be put into cold storage until the War is over. To work for a new order based upon love means taking many risks in a world which does not yet accept its authority.

I have written you of some practical expedients; of the campaign for bringing food relief to the starving peoples of Europe; of the organization of vigilance over such ugly growths as 'Vansittartism' and indiscriminate bombing; of the service performed by a group of young men to the outcasts of society whose isolation they share. But these measures depend for their success upon the constant presentation of the War in its widest perspective by the makers of opinion – teachers, preachers, writers, and speakers. On them we rely for surveys of accurate facts, especially those which explain the characteristics of our enemies, and show how the emergence of aggressive minorities in Germany, Italy, and Japan is bound up with the course of previous history. Unless the people of this country have an elementary idea of that perspective before the War ends, and realize that the oppressors, as well as the oppressed, stand in need of mercy, we shall prove as unfit for whatever share we may have in the making of peace as we proved in 1918.

If the scholars and students within the pacifist movement help even a few to understand that Hitlerism is not the cause of this conflict but the result of grave historical errors, they will have begun that process of enlightenment which alone can arrest the present spiritual decline of the civilized world. The vicious circle of recurrent war will be broken for the first time when one side forgoes its 'right' to demand retribution, and instead offers hope, comfort, support. Whenever generosity has been attempted in politics, it has always paid by disarming the enemy. British chivalry to the conquered French in Canada and the defeated Boers in South Africa brought its own reward. The Rush–Bagot agreement, which established the undefended frontier between Canada and the United States in 1818 after a long period of mutual antagonism, gave two nations a century of security and saved them the wasteful expenditure of millions. The reconciliation between Greece and Turkey, brought about after the last War by the League of Nations, meant that in this War the first people to help the starving Greeks were the Turks. Once the *will* to magnanimity exists, such administrative measures as the collection and distribution of food, and the transfer of refugees and minority populations, are matters of reciprocal organization within the capacity of experts.

There will be much disillusionment after this War, and not only in Germany and Japan. Whatever Britain brings out of the struggle, it cannot be the old unchallenged supremacy, the rule of the 'pukka sahib' over the black, brown, and yellow races whom he despised without realizing, in his complacent blindness, that beneath the silence of their servitude they equally despised him. When fighting ceases the exhausted peoples – roused from their nightmare, appalled by their own actions, filled with revulsion and remorse – will not be healed by attempts to track down criminals, exact punishment, and impose retribution. They will be restored to health only by the patience, the practical sense, and the forgiving wisdom of those who have learned to withhold judgement except when they apply it to themselves.

That will be the moment in which the persecuted, repressed, and ostracized minorities from many countries, who have suffered humiliation in the name of human dignity, will speak and be heard. They can afford to wait and prepare for that time. If today their power is no greater than the power of a small leaven, they are none the less contributing to that supreme spiritual effort which alone can save the new generation from another vindictive peace and a Third World War.

Notes

1 *Cf.* page 10, note 2.
2 April 4th, 1942 (italics mine).
3 See 'An Agreed Report on a Deputation of Pacifist Clergy to the Archbishops of Canterbury and York, Lambeth Palace, Tuesday, June 11th, 1940', published by the Anglican Pacifist Fellowship, 1 Adelaide Street, London, W.C. 2. This report was submitted to the Archbishops and was published with their assent.
4 *Sunday Express*, August 16th, 1942.
5 *Evening Standard*, June 24th, 1942.
6 *Daily Mail*, April 6th, 1942.
7 *Sunday Express*, April 20th, 1942.
8 *The Times*, August 15th, 1942.
9 See Arthur Ponsonby, *Falsehood in Wartime* (Allen & Unwin).
10 April 10th, 1942 (quoted in *Peace News*, May 29th, 1942).
11 A corrective to the stories of Hong Kong atrocities appeared on August 30th, 1942, in a dispatch from Montreal to the *Evening Standard*:

> First letters from Hong Kong prisoners since the colony's fall are being received by joyful families and bring the heartening word: 'We are being treated well.' . . . The letters indicate that the percentage of wounded among those writing is low. Officers report that they are being paid and allowed sports, including volley-ball, ping-pong and soft-ball.

Five hundred letters were mailed from Canadian soldiers.

12 *News Chronicle*, May 2nd, 1941.

10

The Shape of the Future

Today, wherever we go, we hear speculations concerning the future. In a favourite book of my youth, *The Story of an African Farm*, Olive Schreiner, describing the death of her heroine, wrote: 'There is a veil of terrible mist over the face of the Hereafter.' The mist which obscures the period immediately before us here and now is hardly less impenetrable. Even the present is dark and confused. The past alone remains clear to those who can interpret it. Does it tell us anything about this age of chaos in which we are living? Can it give us even a pointer towards the shape of the future?

You will remember how, in the third of these letters, I mentioned the nationalism of Britain and other European countries which began about the time of the Renaissance. We live today in a revolutionary epoch because we are reaching the end of that age of the nation-state, and have not yet evolved the political and economic structure which is to replace it. After the last War – which was really the first stage of this one – an attempt to create it was made in the Covenant of the League of Nations. That experiment failed because the moral standards of European statesmen were not equal to the machinery at their disposal. The bold, imaginative generosity which this country was given so clear an opportunity to exercise required a new type of courage and wisdom which our rulers did not possess.

Nor were they capable of the narrow but effective variety of courage which enabled us to build and hold our Empire despite the vices native to imperialism. Their policy was one of weak provocation, which worked neither for war nor for peace. They kept the vices but let the courage go. The birth-pangs of the new era, which might have been less painful, have therefore taken the acute form of the present War. This is nothing less than a struggle between the liberal democratic ideals of the French Revolution (which Britain, incidentally, opposed), and the counter-revolution symbolized by Hitler and the Nazis.

This counter-revolution, being reactionary, uses reactionary weapons: the weapons of war and tyranny. It cannot be permanently conquered by its own weapons, for, believing in them, it will always handle them better than those who take to them reluctantly. The democracies should never have attempted to oppose it by copying its methods. The only effective way of fighting it is with mental and spiritual weapons, opposing it from without, acting as a leaven which undermines it from within. The real match for Hitler today is Gandhi. Gandhi's methods, being also revolutionary but not reactionary, would finally defeat aggression, however temporarily apparent the victory of militarism might be. But our governmental clique has neither the courage nor the imagination to let him apply them even in his own country, whose liberation from imperialism has become the acid test of British intentions in this War.

If you want to find an age of history comparable to our own for disturbance and disaster, turn back to the fall of the Roman Empire. When the Goths sacked Rome in A.D. 410, the men of those days believed that the light of civilization had been extinguished. Their mood must have been similar to that of Sir Edward Grey, who, too prophetically, saw the lamps going out all over Europe in August 1914. But it was just at the time of Rome's eclipse, as a contributor[1] to *The Friend* has reminded us, that Saint Augustine began to set down his vision of a City of God which would outlast all earthly empires.

Augustine could not know that the fall of Rome would be followed by an age of barbarism; and that in its turn by an age of feudalism; and feudalism by an epoch of industrialism, of which the breakdown would lead to the modern totalitarian state. But he could, and did, achieve two things. He could analyse the causes of the downfall of Rome; and he could describe and assess those eternal values upon which life would build itself anew when the cataclysm was past.

We can follow his example today by asking ourselves three questions, and trying to answer them. First, what were the deep underlying causes of the two Great Wars, as distinct from those more immediate causes which historians call 'occasions'? Secondly, how can we rediscover the abiding values in the midst of catastrophe? Thirdly, how far can we begin, however humbly, to order our lives by these values in such a fashion that, whatever may be the shape of the future, we can influence it for the better and thus help to divert the course of history into new channels? To be sure of the things that matter, to distinguish between the temporal and the eternal – that is the first of all duties for peace-makers today.

In my third letter I touched upon the major causes of this War in trying to show you how the history of modern Germany had tended to put her most aggressive minorities into power. Let us summarize them again.

First, chronologically, came the break-up of the unity of Christendom by the Reformation, which set out to remedy the shortcomings of the medieval Church, but in the process created a number of state churches which substituted the religion of nationalism for the religion of Christ. This development was itself part of the second underlying cause: the rise of the nation-states whose leaders, in their different fashions, preached and practised the doctrine of 'Reason of State'. According to this theory, the collective state possessed a godhead of its own and became a law unto itself. Hence the political world has been rent by mutually incompatible conceptions of national sovereignty, and the economic world by a chaotic scramble to control the earth's resources.

When the earliest nations to be united as states developed an urge for empire-building after the voyages of the Great Discoverers had revealed the existence of rich new territories ripe for exploitation, Germany and Italy came late into the race because of their delayed unification. Like Japan, they found themselves 'Have-Not' Powers, who could increase their territories and markets only by means of aggressive military programmes. These programmes have not unnaturally been carried out by buccaneers and paranoiacs who climbed to leadership by taking advantage of their maladjusted communities. Unlike the smaller and less ambitious countries, these so-called 'Great Powers' added to their economic needs a biological impulse, arising from the pressure of national populations which desired to expand within the confines of their own 'state', or, like Japan, to find space and economic opportunity inside some other country with a higher standard of living.

At the beginning of this War, Britain and France between them owned more than 16 million square miles of the earth's surface, excluding the mandated territories lost by Germany after the Treaty of Versailles. In April 1940, two months before the fall of France, I found myself in Paris on my way home via Lisbon from the United States. Everywhere I saw, on the walls of offices, shops, and railway stations, large maps of the world in which French and British territories were coloured red. Beneath these maps ran the caption: 'We shall win because we are the strongest.' From American friends who left France more recently, I have learned that the remnants of the maps can still be distinguished beneath Nazi notices on the walls of Paris – a forlorn commentary on the fact that great possessions are not a source of security, but of the envy, hatred, and fear which lead nations into war.

The third cause of the two World Wars has been the failure of the capitalist system to give security and a decent life to the common citizen. Its emphasis was placed upon private profit-making; the few pursued material gain at the expense of the many, until the State itself was called upon to support the vested interests of profiteers. This trend has been exaggerated by the technological processes of the nineteenth and twentieth centuries, with their consequence in mass-unemployment and their destruction of human personality by mechanization. It was large-scale unemployment, you may remember, that Hitler was elected to remedy. He accomplished this first by eliminating certain classes, such as Jews, women, and Social Democrats, from the labour market altogether; and secondly by building a great military machine. War always offers the easiest apparent solution for unemployment. Actually, it is only a ruinous postponement, which sacrifices great wealth and vast populations to the wheels of Juggernaut.

This machine, with the terrible capacity for devastation which human inventiveness has given it, brings us to the fourth of our basic causes of war. The swift impetus of modern scientific development might have been used to free the lives of all men from want, loneliness, and insecurity. Instead, it has outpaced their moral ability to control its direction, which is now almost entirely destructive. The desire for power is today strangely allied with a universal impulse towards death. Deep in human nature there seems to lie a perversity which is wholly evil, since it is contrary both to man's material interests and his finer aspirations. It is as though Lucifer, when he fell from Heaven, found the means to insinuate a drop of poison into each of the human souls which God has made throughout His ages of Creation. Man's real war is not against his fellow man, but against that drop of poison in himself. One purpose of the new era struggling so frantically into life must be to find and isolate this elusive poison, just as a germ has to be isolated before the disease which it causes can be cured.

To divert the course of history would mean a reversal of the four major trends which, for all their potentiality for good, have brought such evil results. The restoration of religious unity through the revitalization of the Christian Churches; the substitution of an international authority for national sovereignty and competitive imperialism; the change from capitalism to socialism, already visible in the war-time transformation of industry from private ownership to State control; the development of man's moral nature above the level of his scientific inventiveness: these must be the long-range objectives of any society which hopes for permanence. Chronologically, the last should come first, lest the human race destroy itself before it can turn its attention to the other three.

The history of the past four centuries, sharply mirrored in the story of the last 20 years, has proved that the practice of Christianity and the preservation of nationalism are incompatible. One or the other must go. The protection of national cultures has great value, but to preserve the temporal power of the nation-state is fatal to any Kingdom which is not of this world. Those who believe the dominant State to be more important than a Christian society should put all their energies into defending it; but others who accept the validity of the Gospels cannot defend the State when to maintain it means the sacrifice of the faith by which they live. For them this War is the inevitable Nemesis that has followed the errors made by Allied statesmen after the last. If these errors had not produced their Day of Judgement, the working of a moral law in history would be less clearly demonstrated and Christian promises less abundantly fulfilled. Never has it seemed so incontestable that different results can be obtained only by a full acceptance of Christ's teaching and a sincere attempt to build the Kingdom of Heaven on earth. The sacrifices of power and prestige which that Kingdom demands must be faced and fulfilled before the race of men can count on survival:

> It is as natural [writes a New Zealand clergyman] for the world to be at war today as it is for typhoid germs to breed in open drains and for frogs to breed in stagnant pools ... It is the chastisement of men upon men for the flagrant disregard and criminal distortion of Divine Truth.[2]

Yet the same men can construct, if they desire, a new international organization in which the worst evils of the State will disappear; they can control those vested interests which take so much more out of society than they ever put into it; they can initiate an age of science in which the chief object of study and experiment will no longer be military destruction, but the vast unexplored continent of the human mind where the roots of war lie. These voyages of psychological discovery will perhaps be part of a new education, directed towards producing a finer type of being, both physically and morally. The problem is one of creating a spiritual authority strong enough to discipline mental capacity, which has outrun the power of the soul.

At present these high achievements are objectives only. There will be no historical vacuum in which their pursuit can take place unimpeded by the contrary pressure of circumstances. Even their preparation cannot begin without some attempt to visualize alternative situations and the type of action which these will compel. The character of the future, and the work of those who endeavour to foresee it, must be determined by one of three possible outcomes of the War: an Allied defeat, an Allied victory, or a negotiated peace following a stalemate or a revival of sanity.

You have probably observed that, the larger our catastrophes, the louder become the prophecies of 'ultimate victory'. As Nathaniel Gubbins, the *Sunday Express* humorist, remarked in a 'Party Conversation',[3] 'The worse the news is, the more we talk about what we are going to do with Germany after the War.' Far more reassuring than this universal variant on whistling to keep up our courage are some facts which recently appeared in *Fellowship*,[4] a monthly magazine published by the American Fellowship of Reconciliation. The article in question, 'You Cannot Kill the Spirit', was written by Leonard S. Kenworthy, Head of the Social Studies Department in the Friends Central School at Overbrook, Pennsylvania, who directed the Quaker Centre in Berlin from June 1940 to June 1941. Mr Kenworthy describes the persistence of religion in Germany despite the Nazis, and attaches especial importance to the small groups all over the country who come together for spiritual fellowship. He writes:

> They are not political, they are religious cells. In many respects they resemble the small communities of first-century Christians. They read and pray together. They discuss their personal problems with each other. They are extremely important to the future of Christianity in Central Europe.

Here is a prescription for the revival of religion more reliable than our long-pursued but elusive victory. It offers us a tolerable way of life even if we never catch up with the political and military triumph of which our leaders still feel so certain. In Germany itself, the Nazis have been unable to destroy the spirit. The present methods of the

German Christians might well be our only defence against the dread future suggested by Mr Churchill in a speech made on November 11th, 1938: 'I have always said that, if Great Britain were defeated in war, I hoped we should find a Hitler to lead us back to our rightful position among nations.'[5]

Dark as such a prospect appears, many responsible leaders of opinion are coming to realize that 'ultimate victory' may present us with problems at least as appalling. One of the best recent summaries of its probable consequences was contributed by the Bishop of Chichester to the *Christian News-Letter*.[6] Describing a recent visit to Sweden, the Bishop remarks how 'strangely isolated' England now is from most European countries. Newly enlightened with regard to Europe from the angle of the peoples concerned, the Bishop inquires what will happen in the political vacuum created by the downfall of Hitlerism:

> When the crash comes – *What next*? . . . Even before food can be distributed, order is indispensable. With the collapse of the Nazi regime there is the immediate danger of civil war all over Germany, in which the two million men and women from the Occupied Countries now doing forced labour as slaves may have something to say . . . Once the news reaches Norway, Holland, Belgium, Moravia, Yugo-Slavia, that Hitler and all his men have fallen – *What then*? Just because the provocations have been so great we have to beware of a different kind of bloodshed on a terrible scale . . . The whole situation is big with peril.

In the light of the possibility here suggested that 'the last state of Europe may be worse than the first', what rational person can regard 'victory' as something to be cheerfully acclaimed and pictured in glowing colours, however long its achievement may take? Modern civilization, with its large populations which have to be fed and its crowded industrial cities where epidemics spread like forest fires, cannot afford universal war. The cost in human suffering is too great, the breakdown of the machinery of living too complete. Until the present conflict ends, our work for the diversion of history into new channels can be no more than a preliminary looking forward, an uncertain preparation.

Hence the most practical move we can make is to press for the ending of the War by an Armistice, a prolonged Conference, and a negotiated Peace. 'There is no way of ending sin except persuading sinners to leave it off,' writes Walter Walsh in *Jesus; War or Peace?* There is no way of ending war except by urging those who are fighting to give it up, and try a better method of settling their dispute. Even a pause might recall how little connection exists between its long-ago 'occasions', and those fundamental causes which war always aggravates and can never remove.

Meanwhile, we have to confront and try to reduce our own share of responsibility for humanity's failure. What really underlies the sorry story of four centuries, the deeper causes of revolutionary war traced from the Reformation to the present day? Is not the same tale of greed, ambition, fear, and vindictiveness to be found in our own lives and our relations with our neighbours? The acquisitiveness exposed by rationing, the widespread hoarding of scarce commodities, the envy of a neighbour's house, the secret joy in an acquaintance's downfall: are they not evidence of the same self-interest which when enlarged becomes capitalism, imperialism, and competitive nationalism? We shall make the shape of the future no better than the ugly shape of the past unless we learn to give more than lip-service to the Christian virtues which most of us extol. The failure has not lain in the political machinery at the disposal of the age, but in those who misused or ignored it. 'We needs must love the highest when we see it,' wrote Tennyson; yet man, offered the highest, has deliberately permitted his lower nature to occupy the driver's seat.

But not always. Most of our everyday social rules are based upon the abundantly justified assumption of decency, kindness, and common sense in the mass of human beings.[7] Our traffic laws rest on the supposition that travellers will respect one another's safety, and refrain from running somebody down the moment that they take to the road. Our postal system is founded on the well-established belief that the great majority of mankind can be trusted with other people's possessions. When ordinary men and women from different nations get together for some international purpose, they do not instinctively dislike one another. They begin to hate only when war-making politicians inflame their emotions with propaganda, and when their rulers, corrupted by power, urge them to deeds of violence. Search through the private correspondence of friends, or the small items on the inner pages of obscure magazines, and even now you will discover numerous examples of kindness on the battlefield, in prison, in hospital, or internment camp. The best hope for the future lies in the thousand similar instances of human cooperation that we shall find when the warring governments lift the veil of silence through which only stories of cruelty and hatred are permitted to emerge.

Some day peace will return, and with it, perhaps, a measure of sanity. Man, though foolish, lethargic, and corruptible, has nevertheless an undying capacity for spiritual resurrection; his impulse towards death is balanced by an indestructible principle of life. In that hour of awakening men and women will examine, realistically and with horror, the consequences of this madness of destruction which their leaders began and which they accepted; and will turn, as they turned once before, to the few who kept their heads and provided a nucleus of reason in a world of confusion.

When the pacifist minority is no longer penalized by the wielders of power, it may well be asked to consider how the common peoples of Britain and Germany, purged by suffering, may find a way of life together, freed from the menace of their privileged classes to their mutual interests. Perhaps we shall help our ex-enemies to acknowledge their acquiescence in evil by admitting our own; for we in this country have also a confession to make. We too have propounded, in our colonial territories and above all in India, an intolerable theory of racial superiority; and if we did not seek to 'dominate the world', it was because, in markets and on trade routes, we were supreme already. We, the British people, consented to the dividing of Europe in our own interests, instead of uniting it for humanity's welfare. We allowed our leaders to scheme for power, instead of creating an international society based on confidence and friendship. For our own lack of vision, the common peoples of the world have paid. Perhaps the weary days spent in food queues and the sleepless nights passed in shelters will acquire a retrospective value from the mutual sympathy felt by the men and women of many nations who have endured both; and, understanding that our fears, our hopes, and our affections are the same, we shall turn to one another for much-needed cooperation in rebuilding the beauty that has been destroyed.

That work of reconstruction will demand every gift which can be contributed to the common weal by all the races of men: the black, the brown, the yellow, the white, 'whose prayers go up to one God under different names'. Let us therefore seek to end the unnecessary handicaps which have hitherto prevented the full realization of such gifts: the race and class prejudice that sets artificial barriers to the progress of millions; the frustration of women's abilities by tradition and self-interest; the tyranny which age still exercises over youth. The mobilization of all the capacity available on earth will not be too great for the restoration of order to the chaos which this conflict will leave behind it, nor for the solution of the human problems upon which the shape of the future depends.

Notes

1 F. J. Tritton, March 13th, 1942.
2 The Rev. C. W. Chandler, in the *Auckland Star*, January 31st, 1942.
3 *Sunday Express*, July 5th, 1942.
4 June 1942.
5 See also his book *Great Contemporaries*.
6 June 24th, 1942.
7 The daily newspapers are filled with evidence that man's instinct is to save life, not to destroy it. For example; the *Daily Herald* for August 24th, 1942, contained a striking description by Hannen Swaffer of the rescue at Ponty-pridd of a child evacuee, who had fallen down a mountain crevice in land that was in danger of subsidence, after 17 men and boys had struggled unceasingly to reach her for 15 hours.

Epilogue

My Dear Son, Ten letters and a hundred pages have not, after all, been sufficient for the statement of faith that I meant to give you. Perhaps a thousand pages would still not be enough. Or perhaps everything I wanted to say is summed up in two quotations which I have discovered in the last few days.

One comes from the broadcast of a Canadian historian[1] who spoke to the British people from London in the winter of 1941. These were his words:

> We have watched the plain, ordinary men and women of Britain forget the grievances that make men impatient in the griefs that make them patient . . . We have both been purged by great pities and many mercies. The only race we wish to see triumphant on earth is the human race.

The other I found in the weekly newsletter published by Dr Josiah Oldham.[2] He writes:

> The secret of the power of Christianity is that it does not shrink from the abyss . . . We are made aware in hours of defeat that life has dimensions of which we too easily lose sight when everything is going well.

Here, at any rate, is my own summary of all that I have tried to say to you in the past few weeks.

I believe that men cannot fully exercise compassion until they have experienced humiliation.

I believe that, before they can help the pariahs and felons of society, they must stand beside them in the dock and the prison.

Just as the fight for personal self-conquest takes place in each new generation, so in every epoch has been waged the struggle to eliminate war. But that conflict has now reached a point at which man must overcome this worst of his enemies or himself be annihilated.

War begins first in the human soul. When a man has learned how to wrest honour from humiliation, his mastery of his own soul has begun, and by just that much he has brought the war against war nearer to victory. He no longer looks to the men on the heights to supply him with evidence that God exists. Instead he himself, from the depths, becomes part of that evidence. He has proved that power itself is powerless against the authority of love.

That you and I will each leave some such evidence behind us, is the perpetual hope of

Your Mother

Notes

1 Leonard W. Brockington, on December 22nd, 1941.
2 *Christian News-Letter*, July 15th, 1942.

Original cover of *Seed of Chaos: What Mass Bombing Really Means*, first published in 1941 for the Bombing Restriction Committee by New Vision Publishing Company, London.

Seed of Chaos

What Mass Bombing Really Means

Vera Brittain

Then rose the seed of Chaos, and of Night, To blot out order and extinguish light.

Pope, The Dunciad, *Book IV, L.13*

Were they ashamed when they had committed abomination?

Jeremiah 6, 15

Sleep No More

Do they bear pity in their planes,
Or any doubt of the deed?
Must the machine-bewildered mind
Beat on in ruthless speed?
Are cries too faint and far below
For flyers to give heed?
The bitter groan of Hamburg streets,
The weeping in Milan,
Death-cries of children trapped and burned –
Oh, justify who can!
For they pierce like a terrible sword
Through the heart of an Englishman.
O lovely, brave young flyers!
What judgement shall we receive
Who sent you out in the dark with death
While we sleep through the night and live?
When we wake from our deeper dreaming
Will you pity us? And forgive?

W.R.H.

I

Introduction

We British are not an imaginative people. Too often the value of our genuine kindliness is nullified by a native inability to picture the effects of our actions upon other nations, or even upon unfamiliar sections of our own community. Throughout our history wrongs have been committed, or evils gone too long unremedied, simply because we did not perceive the real meaning of the suffering which we had caused or failed to mitigate. Whenever we have actually understood the pain and misery awaiting alleviation, as in the case of slavery, or of child labour in the mines and factories of this country during the nineteenth century, we have been the first to rise and demand a change of policy on the part of our rulers.

The purpose of this book is to inquire how far the British people understand and approve of the policy of 'obliteration bombing' now being inflicted upon the civilians of enemy and enemy-occupied countries (including numbers of young children born since the outbreak of war) by ourselves and the United States. The propagandist Press descriptions of this bombing and its results skilfully conceal their real meaning from the normally unimaginative reader by such carefully chosen phrases as 'softening-up' an area, 'neutralizing the target', 'area bombing', 'saturating the defences', and 'blanketing an industrial district'.[1]

Up until the summer of 1943, the short paragraphs which gave more factual descriptions of our raids and their consequences were apt to appear in small print on the back pages of newspapers or at the foot of main columns. Since that time one or two newspapers, and particularly the *Daily Telegraph*, have begun to report the R.A.F. bombing in a franker and more conspicuous fashion, and even to carry special articles by air experts or pilots, while the gigantic raids on Berlin which began in November, 1943, were apparently treated as gala occasions on which the whole Press was permitted to let itself go. But it is only when the facts are collected, and the terrible sum of suffering which they describe is estimated as a whole, that we realize that, owing to the R.A.F. raids, thousands of helpless and innocent people in German, Italian and German-occupied cities are being subjected to agonizing forms of death and injury comparable to the worst tortures of the Middle Ages.

From the extreme discomfort of this realization, the British citizen seeks to escape by two main arguments. In the first place, he maintains, mass bombing will 'shorten the War' – a contention now much favoured by Ministers, officials, Members of Parliament, and some leading Churchmen.

In reply it can justly be stated, first, that there is no *certainty* that such a shortening of the War will result; and that nothing less than absolute certainty entitles even the most ardent of the War's supporters to use these dreadful expedients, which an informed commentator, quoted in the *New Statesman* for December 18th, 1943, described as 'British boys being burnt to death in aeroplanes while they are roasting to death the population down below': Mr Churchill himself has called the mass bombing of German cities an 'experiment'.[2] What does appear certain is the downward spiral in moral values, ending in the deepest abysses of the human spirit, to which this argument leads. Those who remember the first Great War will recall that precisely the same excuse – that it would 'shorten' the period of hostilities – was given by the Germans for their policy of *Schrecklichkeit* (terror), and was used in connection with their submarine campaign. We ourselves refused to accept the argument as valid when the Nazis revived it in this War to justify the bombing of Warsaw, Rotterdam, Belgrade, London, and Coventry.

Secondly, when the word 'shorten' is used, it generally implies the limiting or reduction of the total amount of human suffering and destruction. Such a time test or standard is misleading. In a vast, concentrated raid, lasting a few minutes, more persons may be killed or injured than in a modern major battle lasting two or three weeks, in addition to the destruction of an irreplaceable cultural heritage of monuments, art treasures, and documents, representing centuries of man's creative endeavour. In fact, the mass bombing of great centres of population means *a speed-up of human slaughter, misery and material destruction superimposed on that of the military fighting fronts*.

Thirdly, the 'experiment' has demonstrated, so far, that mass bombing does not induce revolt or break morale. The victims are stunned, exhausted, apathetic, absorbed in the immediate tasks of finding food and shelter. But when they recover, who can doubt that there will be, among the majority at any rate, the desire for revenge and a hardening process – even if, for a time, it may be subdued by fear? Thus we are steadily creating in Europe the psychological foundations for a Third World War.

Fourthly, the systematic destruction of German industrial power is hitting ourselves from a long-term standpoint. The prosperity of the Continent in the past has largely depended upon German industry, and the continental market, including Germany, was the best block of markets for the British export trade before the War. Mr Harcourt Johnstone, Secretary of the Department of Overseas Trade, said on 24th November, 1943, that in order to maintain our standard of life after the War, we should need to increase our export trade by £350 million a year. The destruction of German industry means the destruction of the biggest British export market, and makes it more difficult, if not impossible, to prevent the lowering of the standard of life in Great Britain for many years after the War.

The second main argument brought forward to excuse our present policy of obliteration bombing is that we too have suffered – as indeed we have – and that therefore we are fully entitled to pay back what we have endured.

Mr George Bernard Shaw has dealt characteristically and drastically with this double contention in a recent letter to the *Sunday Express* (November 28th, 1943):

> The blitzing of the cities has carried war this time to such a climax of infernal atrocity that all recriminations on that score are ridiculous. The Germans will have as big a bill of atrocities against us as we against them if we take them into an impartial international court.

There are three further replies which should also be carefully considered by all rational people.

In the first place, investigations into the origins of *civilian* bombing far from the fighting lines (as distinct from the bombing which forms part of a military campaign) make clear the difficulty of justly assessing with whom lays the fault of starting it. The cumulative growth of civilian bombing to its present nightmare stage seems on present information to be an outstanding instance of the tragic fashion in which war-time cruelty grows like a snowball by its own momentum once the power of Juggernaut has taken control. Some accidental violation of international law, assumed to be deliberate, is repaid by a reprisal 'in kind'. The enemy 'hits back'; we retaliate harder still; in each case the accidental consequences (such as the bombing of a church in mistake for a factory) are advertised for propaganda purposes by the victim as having been intentional. So the grim competition goes on, until the mass-murder of civilians becomes part of our policy – a descent into barbarism which we should have contemplated with horror in 1939.

Secondly, it is a fact – as this book will show – that though parts of Britain suffered cruelly in the Blitz, some of the terrible inventions and tactics now being used were not known or practised at that stage of the War. Even in those early days, the knowledge of our distress and confusion was limited to the areas which endured them, and particularly to the surviving victims and to Civil Defence and rescue workers, who had actually to deal with the shambles to which German bombs reduced many humble homes. It is, I believe, the comparative rarity of first-hand experience among the majority of the British people which accounts for their supine acquiescence of obliteration bombing as a policy.[3]

My own experience is relatively small, but as a Londoner who has been through about 600 raid periods and has spent 18 months as a volunteer fireguard, I have seen and heard enough to know that I at least must vehemently protest when this obscenity of terror and mutilation is inflicted upon the helpless civilians of another country. Nor do I believe that the majority of our airmen who are persuaded that mass bombing reduces the period of their peril really want to preserve their own lives by sacrificing German women and babies, any more than our soldiers would go into action using 'enemy' mothers and children as a screen.

In the third place, retaliation 'in kind' and worse means the reduction of ourselves to the level of our opponents, whose perverted values have induced us to fight. However anxious we may be to win the War, the way in which we win it will also determine our future standing as a nation. If we imitate and intensify the enemy's methods, we shall actually have been defeated by the very evils which we believe ourselves to be fighting.

It is to the credit of some of the worst-bombed areas of this country that many of their inhabitants have recognized this important truth. In April 1941, when the British Institute of Public Opinion carried out a survey of the whole country's response to the question: 'Would you approve or disapprove if the R.A.F. adopted a policy of bombing the civilian population in Germany?', it was noticeable that the people of the heavily bombed areas were less in favour of reprisal bombing than those who had escaped the raids. The largest vote in favour of reprisals (76 per cent of the population) came from the safe areas of Cumberland, Westmorland, and the North Riding. In the bombed districts of Inner London, which had then endured eight months of heavy and continuous raids, 47 per cent disapproved of reprisals, 45 per cent approved, and the rest were undecided (*News-Chronicle*, May 2nd, 1941).

When, therefore, on July 15th, 1941, Mr Churchill said at the County Hall: 'If tonight the people of London were asked to cast their vote whether a convention should be entered into to stop the bombing of all cities, the overwhelming majority would cry, "No, we will mete out to the Germans the measure and more than the measure, that they have meted out to us",' he was disregarding opinions ascertained only two months earlier.

'I wouldn't wish this trouble on any other woman!' cries the young mother in A. Burton Cooper's Lancashire play, *We Are The People*, after her small boy has been blown to pieces by a daytime bomb on a local playground. And that, I believe, is the normal reaction of every decent person, once real knowledge has come to him or her through individual suffering.

It is because I want you, the readers of this book, to have such knowledge, so far as facts ascertained from sources available under war-time conditions can give it to you, that I am going to describe, with references to my sources of information, what our bombing policy means to those who have to endure its results. I shall have to quote some horrible details, but these are not included from sensational motives. They are given in order that you who read may realize exactly what the citizens of one Christian country are doing to the men, women, and children of another. Only when you know these facts are you in a position to say whether or not you approve. If you do not approve, it is for you to make known your objection – *remembering always that it is the infliction of suffering, far more than its endurance, which morally damages the soul of a nation.*

Notes

1 The use of soporific words to soothe or divert the natural human emotions of horror and pity is a characteristic and disturbing feature of this War.

2 See quotation from *Time*, page 118, note 2.

3 These conclusions are strongly reinforced by the conversation reported on December 18th, 1943, between 'Critic' of the *New Statesman* and the informed commentator already mentioned:

> My friend asked, 'Do you think people's consciences are much roused about the bombing we are doing now?' I replied that a few papers and some people gloat, but most people uncomfortably acquiesce in the absence of any alternative. They have got used to the idea and don't know many details. 'Well,' said my friend, 'it is certainly very odd. They worried a good deal when they thought we could be hit back and we were ourselves doing comparatively little damage, but today our raids are of quite another order. We drop liquid fire on these cities and literally roast the populations to death.' 'Critic' went on to point out the 'brainless' and purely destructive character of this policy, since it is accompanied by no kind of political appeal to the anti-Hitler elements in the suffering populations.

2

The History of our Bombing Offensive

I Changes of Policy

In the early days of the War, the Government's insistence that Britain bombed only military objectives was matched by the righteous indignation of official *communiqués* whenever German bombs fell on churches, hospitals, schools, or private dwellings. From the highest ministerial circles to the lowest, this scrupulousness was maintained. On April 17th, 1943, a letter from Sir Leo Chiozza Money to the *New Statesman* concluded with these words:

> Swift is the growth of hate. I would like to remind your readers that it was as recently as January 27th, 1940, that Mr Winston Churchill, then First Lord of the Admiralty, condemned the bombing of an enemy as a 'new and odious form of attack', and refused to listen to the clamour then arising that bombers should cease to drop leaflets by night and load up with 'beautiful bombs'. I may add that when the present writer read Mr Churchill's strong denunciation of a new and odious form of warfare, it seemed to me (I wrote on February 10th, 1940) that I recognized the good sense and good feeling I had always associated with the man who uttered it.

It is necessary only to quote some recent ministerial pronouncements to realize how violent has been the change in the attitude of the Prime Minister and his deputies during the past 18 months, and how steep the decline in moral standards since the opening days of the War.

On June 2nd, 1942, Mr Churchill gave the following undertaking to the House of Commons:

> As the year advances, German cities, harbours, and centres of war production will be subjected to an ordeal the like of which has never been experienced by any country in continuity, severity, or magnitude.

Nearly a year later, addressing Congress at Washington in May, 1943 (*News-Chronicle*, May 20th), he issued a similar threat to Japan:

> It is the duty of those who are charged with the direction of the War to . . . begin the process so necessary and desirable of laying the cities and the other military centres of Japan in ashes, for in ashes they must surely lie before peace comes back to the world.

Although *The Times* correspondent stated (May 20th, 1943) that this promise 'found ordinary applause insufficient and Members shouted their approval', it seems possible that the American reaction to his obliteration policy was not precisely what the Prime Minister expected: on September 8th, 1943, the *News-Chronicle* reported President Roosevelt as assuring Congress that 'We [i.e. the Americans] were not bombing tenements for the sadistic pleasure of killing as the Nazis did, but blowing to bits carefully selected targets – factories, shipyards, munition dumps.'

Nevertheless, having spoken in July of 'the systematic shattering of German cities', Mr Churchill further expanded his theme to the House of Commons on September 21st, 1943:

> The almost total systematic destruction of many of the centres of German war effort continues on a greater scale and at a greater pace. The havoc wrought is indescribable and the effect upon the German war production in all its forms . . . is matched by that wrought upon the life and economy of the whole of that guilty organization . . .

On the same occasion he told the House: 'There are no sacrifices we will not make, no lengths in violence to which we will not go to destroy Nazi tyranny and Prussian militarism.' A week or two later, in a message to Bomber Command, he described this process as 'beating the life out of Germany'.

It is interesting to compare these savage threats with one of the concluding paragraphs in Mr Churchill's own book, *The World Crisis*, first published in March 1929, and dedicated 'To All Who Hope':

> The disproportion between the quarrels of nations and the suffering which fighting out those quarrels involves; the poor and barren prizes which reward sublime endeavour on the battlefield; the fleeting triumphs of war; the long, slow rebuilding; the awful risks so hardily run; the doom missed by a hair's breadth, by the spin of a coin, by the accident of an accident – all this should make the prevention of another great war the main preoccupation of mankind. No more may Alexander, Caesar and Napoleon lead armies to victory, ride their

> horses on the field of battle sharing the perils of their soldiers and deciding the fate of empires by the resolves and gestures of a few intense hours. For the future they will sit surrounded by clerks in offices, as safe, as quiet and as dreary as Government departments, while the fighting men in scores of thousands are slaughtered or stifled over the telephone by machinery. We have seen the last of the great Commanders. Perhaps they were extinct before Armageddon began. *Next time the competition may be to kill women and children, and the civil population generally, and victory will give herself in sorry nuptials to the spectacled hero who organizes it on the largest scale.* (Italics mine)

It is hardly surprising that Mr Churchill's present subordinates have followed his lead with imitative threats. At Quebec in August, Mr Brendan Bracken, Minister of Information, stated to the Press: 'Our plans are to bomb, burn, and ruthlessly destroy in every way available to us the people responsible for creating this War' (*News-Chronicle*, August 20th, 1943). On October 13th, 1943, in a speech in London at the Constitutional Club, Sir Archibald Sinclair, Secretary of State for Air, declared: 'Nor does Germany yet know the worst. We are not near the culmination of the offensive . . . Make no doubt about it, their wounds will prove mortal.' On November 5th, at a speech in Cheltenham, he added: 'We shall continue to hammer the enemy from the skies till we have paralysed their war industries, disrupted their transport system, and broken their will to war.'

Again, at Plymouth on January 22nd, 1944, he announced:

> Our aim is to paralyse German war industry and transport, and our objectives are not cathedral towns, but cities which are the centres of German war industry and transport, and nothing will divert us from our aim.

If Aachen, Cologne, Munich, Münster, Mainz, and Magdeburg are not cathedral cities, what are they? As for Lübeck, Nuremberg, Frankfurt, and Leipzig, they were treasure-houses of art and history, comparable with Salisbury, Oxford, Chester, and Edinburgh. Are we to suppose that Germany is a country unknown to Sir Archibald – or that he is counting on British ignorance of German cities?

Echoes of these threats have appeared in the Press in articles and paragraphs so numerous that it is impossible to quote more than a few. In the *Sunday Express* of April 20th, 1942, the new policy was welcomed by Mr John Gordon, the Editor, in words of which the implications become more astonishing the longer they are considered:

> Germany, the originator of war by air terror, is now finding that terror recoiling on herself with an intensity that even Hitler in his most sadistic dreams never thought possible.

On June 1st, 1942, a B.B.C. broadcast made it clear that our own 'sadistic dreams' did not exclude the death and injury of children:

> We are even sorry for the women and children who may have suffered for the stupidity of their menfolk in putting Hitler in power and their cowardice in keeping him there. We are sorry for them, but when we remember Warsaw, Rotterdam, Coventry, and Belgrade, and wonder how many of the women and even the children of Cologne exulted at the activities of the Luftwaffe, then we harden our hearts. For the German women, and even the German children, ought to be capable of recognizing evil and rejecting it.

One may perhaps inquire whether the speaker remembered that a large number of the children now being burned and mutilated in Germany have been born since the outbreak of war. In what sense can these babies be said to have 'exulted' at the activities of the Luftwaffe, or to be capable of recognizing and rejecting evil?

The following day, however, the *News-Chronicle* reinforced this broadcast with its own comments:

> The German people must be made to feel in their own bricks and bones the mad meaning of their rulers' creed of cruelty and destruction . . . If by the ferocity of our retribution we can convince them at last that violence does not pay and induce them to become good citizens of the world – then the loss of their monuments will be as nothing compared to the contribution to our common inheritance which their conversion to civilized conduct will make.

The idea that 'ferocity' and 'retribution' make people into good citizens and convert them to civilized conduct goes strangely contrary to the scientific findings of psychology and the experience of penal reformers during the past 150 years. It also violates the most elementary principles of common sense. A people who have been beaten into apathy and defeatism by their conquerors are hardly likely to conclude that violence does not pay.

By August 1943, shortly after Hamburg had suffered an ordeal probably more appalling than any to which a city of comparable size has ever been subjected (see section on Hamburg, p. 132), our threats had become more specific. On August 19th, various newspapers quoted a British United Press report from Quebec: 'At least 50 of Germany's main cities will meet the fate of Hamburg by Christmas.' Two months later, on October 13th, an article by Group Captain Hugh Edwards, V.C., D.S.O., D.F.C., in the *Daily Mail* made it clear that this policy was being carried out: 'Bomber Command is obliterating vast areas of industrial Germany.' He remarked, somewhat naively: 'We in Bomber Command sometimes wonder if people realize this,' and added: 'It is certain that people inside Germany are aware of it with grim reality.'

2 Changes of Method

The change from 'precision' to 'obliteration' bombing has necessarily involved extensive changes in method and tactics. Though the policy of the Government is still officially the bombing of military objectives, these objectives have been extended to cover the 'hearts' of great cities, containing offices, flats, tenements, and workers' dwellings; any 'target', in fact, that will destroy Germany's morale and break the will of her people.

This was admitted in the House of Commons by Sir Archibald Sinclair on March 31st, 1943, in reply to a question by Mr R. R. Stokes, who asked the Secretary of State for Air whether on any occasion instructions had been given to British airmen to engage in area bombing rather than to limit their attention to purely military targets. Sir Archibald replied: 'The targets of Bomber Command are always military, but night bombing of military objectives necessarily involves bombing the area in which they are situated.'

Bombing an area in which military objectives are situated is, of course, exactly the method for which we condemned the Nazis in Warsaw, Rotterdam, and Belgrade. The lack of compunction to which it leads is illustrated by the following comment in the *Sunday Dispatch* of March 21st, 1943:

> Bomber personnel, often in miserable weather, and under attack by vicious fighters, try to hit their targets. Any attempt to persuade them to worry unduly about civilians is an attempt to impair their military value.

The same loss of scrupulousness is shown by the actual carrying out of raids under conditions in which it is not even possible to distinguish residential areas from military or industrial targets. This growing practice was carried to extremes in a raid on Frankfurt on November 25th, 1943. According to the *Daily Herald* of November 27th, 'there was thick cloud three miles deep' over the city when the bombers arrived: 'The crews saw nothing of the town. It was blind bombing.' The same process of bombing through heavy clouds was adopted during the raid on Cologne at the end of June, 1943, during which the cathedral was damaged.

On July 27th, 1943, the *News-Chronicle* quoted a 'Military Correspondent' as follows: 'Victory is won behind the lines by the demoralization of the civil population like the spread of fire in the wind.' This appears to convey an official determination to make deliberate war on civilians. The purpose of such a policy had previously been put into words, also in the *News-Chronicle*, by a 'Foreign Observer' on May 17th, 1942:

> I understand that the principle behind your new mass raids is to make sure of hitting important targets *by wiping out the whole area in which they lie*, instead of trying to pick them out one by one. This is obviously the only way to get results in a highly industrialized country like Germany, but it also has the advantage of producing an automatic effect on the population . . . There is a definite point at which the number and weight of bombs dropped in a certain area within a certain time produces no longer a fitful feeling of alarm among those who happen to get bombs dropped near them, but an unbearable strain, more or less approaching panic, on the minds of all the people in that area.

3 Changes of Tactics

The adoption of area bombing has been marked by many new developments which have increased the terror and torture of our attacks. The chief of these has been the use of 'cascade bombing' – otherwise known as 'saturation raids' – by which a great number of heavy bombs are dropped on a limited area in a period so brief that immense destruction goes on simultaneously in all parts of the target city, and the defences are unable to function effectively. The method is to set the centres of cities on fire by means of many thousands of incendiary bombs, after which later planes dump successive loads of high explosive on fires already started – of necessity indiscriminately, since visibility is obscured by smoke.

On July 30th, 1943, this comment appeared in *The Spectator*:

> Such a phenomenon as the discharge of 2,300 tons of explosives and incendiaries over a limited built-up area within 50 minutes has no sort of parallel in history. The heaviest of the raids on London, terrible as they seemed to us at the time, were by comparison quite small affairs.

A recent Press photograph showed a row of 4,000lb 'block-buster' bombs, which take two or three days to fill with high explosives. Four people can stand inside the casing of these huge bombs.[1] In the *Daily Telegraph*, July 17th, 1943, a message from Washington forecast the construction of bombs 'more than twice the size and with many times the destructive power of today's four-tonners'.[2] United States air experts then stated that 'air power on a scale which would make the worst nights Germany had ever suffered seem almost trivial would be thrown into the Battle of Europe before the end of the summer'. According to the same message, advocates of bigger bombers claimed: '5,000 ton bombing raids could be carried out within a matter of months. The destructive force of such a volume composed of the latest bombs could only be conjectured.'

A paragraph in the *Evening Standard*, September 2nd, 1943, disclosed that American super-bombers, capable of carrying these tremendous bomb-loads, are already being built in Paterson, New Jersey, by the Wright Aeronautical Corporation. General Arnold, Chief of the U.S. Army Air Force, stated that these flying battleships would be able to deliver 'massive loads of devastating destruction' and fly the Atlantic and back without refuelling. The comment of the *News-Chronicle* on July 3rd, 1943 – 'Development of the R.A.F.'s assaults to the scale where a large industrial city can be virtually wiped out in a single night have been extraordinarily rapid' – is already an understatement of the preparations which civilization is making to annihilate itself.

On July 27th, the largest number of bombs to be dropped in the shortest period were used on Hamburg in an attack described in the Press as 'the heaviest air raid of all time':

> A million fire-bombs and hundreds of huge two-ton 'block-busters' were dropped in 45 minutes, five minutes quicker than in the 2,300-ton raid on the same target on Saturday. Every such cut in the

> bombing period means greater destruction and greater safety for men and aircraft. Defences are swamped.

During this sequence of raids, according to a paragraph in the *Daily Telegraph*, August 12th, 1943, the Germans reported a new R.A.F. method of 'swamping the defences':

> The R.A.F., they state, at the beginning of a raid mark the target area by a series of 'rings' of green flares, and the following waves of bombers drop their bombs in the periphery of these rings, so as to cut off the German A.R.P. personnel to prevent it reaching these districts. Then the interior rings are littered with bombs and incendiaries.
>
> The terrific heat causes a vacuum of air in the bombed districts, and air rushes from other parts of the town. In this way regular tornadoes arise. They are so strong that people are thrown flat on the ground, and the fire brigades cannot get to the blitzed area with their equipment.
>
> These violent currents of air serve to spread the fire to surrounding districts.
>
> The 'ring-system', which was first used over Hamburg, has another effect. A number of people there died through lack of oxygen caused by the terrible heat.
>
> Hamburg has excellent shelters; they are, in fact, real bunkers, but it was found on opening some that though they were undamaged, many people had died from suffocation.

When these Hamburg raids were concluded, Ronald Walker, Air Correspondent to the *News-Chronicle*, reported:

> The six attacks on the city, port and U-boat yards of Hamburg during four nights and three days probably come nearer than any other series of attacks on Germany to the Harris aim of blotting out a target . . . Air Chief Marshal Sir Arthur Harris, R.A.F. Bomber Command Chief, has made his bombing plan quite plain – the complete destruction of the German industrial cities and ports, one by one. His ideal is to pound them with blows of devastating weight and to keep up that pounding until there is no question of salvage or repair.

4 Sir Arthur Harris and His Strategy

The change of R.A.F. policy from 'precision' to 'area' bombing began on March 3rd, 1942, with the appointment of Sir Arthur Travers Harris to the control of Bomber Command. According to an Associated Press report (March 29th, 1943), Air Marshall Billy Bishop, V.C. –

who told New Yorkers that he did not care 'if there is not one house left standing in Germany' – commended Sir Arthur Harris as 'a tiger with no mercy in his heart' towards the enemy. In the *News-Chronicle* of November 22nd, 1943, Ronald Walker refers to 'the offensive weapon which he more than any other man has forged', and adds: 'In previous articles I have stated that Sir Arthur Harris is out to destroy Germany's key cities section by section.'

This pitiless policy arises, of course, from the determination of our political leaders to 'bomb, burn, and ruthlessly destroy', as Brendan Bracken put it: or, as Churchill expressed it in the House of Commons on July 27th, 1943, that 'in the next few months, Italy will be seared and scarred and blackened from one end to the other'. To a handful of individuals invested with the disproportionate powers conferred by totalitarian war, millions of Germans, Italians, and French owe the devastation of beautiful historic towns, and thousands of families in enemy and occupied countries the death, injury, or mental derangement of young, helpless, and cherished members. These memories alone, of grief and unspeakable horror, are likely to prove an implacable obstacle to the building of a better world.

The old historic towns of Lübeck and Rostock were the first to suffer, in March and April 1942, from the new form of bombing. According to *The Times* of January 1st, 1943, the R.A.F. destroyed more than 40 per cent of the one and 70 per cent of the other. Official photographs show that the centres of these towns were devastated, with the main shopping streets having been the targets, though they were far distant from any recognizable military objectives. Then, on May 30–31st, came the first thousand-bomber raid on Cologne, in which a neutral report gave the number of persons killed in the one night as 20,000, while Abetz, the Nazi representative in Paris, acknowledged between 11,000 and 15,000 dead.

Under the heading, 'High Road to Hell', the American news magazine *Time*, for July 7th, 1943, commented thus:

> The air offensive against Germany and Axis Europe is suffering from understatement. The objective is not merely to destroy cities, industries, human beings and the human spirit on a scale never before attempted by air action. The objective is to defeat Hitler with bombs, and to do it in 1943.

Is the 'understatement' referred to by *Time* perhaps due to a recognition, by those who are responsible for the R.A.F. onslaught, that the ordinary decent citizens of Britain and the USA would not continue to acquiesce in this type of bombing offensive if they were given full details, and realized what these attacks mean for human flesh and blood?

The same issue of *Time* names the men who initiated the change in bombing tactics as 'Air Chief Marshal Sir Arthur Travers Harris, chief of the R.A.F. Bomber Command, and Major General Ira Clarence Eaker, commander of the U.S. Eighth Air Force', but a few lines later the magazine admits, regarding the bombing offensive over Germany, that 'if that knockout is delivered from the air this year, its chief author will be Air Marshal Harris. Except perhaps in the last rounds, the chief instrument will be his Bomber Command.'

Reference is made by *Time* to the Harris broadcast to Germany on July 28th, 1942, in which Sir Arthur made various predictions which are now in process of fulfilment:

> We are going to scourge the Third Reich from end to end . . . We are bombing Germany, city by city and ever more terribly, in order to make it impossible for you to go on with the War. That is our objective. We shall pursue it remorselessly . . . I will speak frankly to you about whether we bomb single military targets or whole cities. Obviously we prefer to hit factories, shipyards and railways. It damages Hitler's war machine most. But those people who work in these plants live close to them. Therefore, we hit your houses and you.[3]

Time precedes its reference to the broadcast by a fragment of Harris' biography, in which Sir Arthur is alleged to be the R.A.F. Commander responsible for developing the 'pacification by bombing' policy which 'kept unruly Indian tribes more or less under control'. Those who recall the Disarmament Conference of 1932 will remember that it was largely in order to retain this method of intimidating the rebellious subjects of the King-Emperor that Sir John Simon and Lord Londonderry 'resisted and obstructed' the proposal to abolish all bombing aeroplanes that was put forward by Italy, and supported by Germany, Russia, and the United States.

It is to the reputed author of this policy that we, the British people, have given the right to take measures perhaps more likely than any others now being used to lose us the peace, and to create in Germany a series of national memories guaranteed to produce a Third World War. We are entitled to inquire how far the Cabinet itself is united in support of these merciless attacks upon helpless civilians.

Some readers may remember a remarkable novel entitled *Public Faces*, published in 1932 by Mr Harold Nicolson (recently Parliamentary Under-Secretary to the Ministry of Information), which went through five editions in less than a year. Its main theme was the transformation of a delicate diplomatic tangle into a dangerous war situation by the irresponsible determination of a stupid and bellicose Air Minister, Sir Charles Pantry, to conduct trial experiments with rocket aeroplanes and an 'atomic bomb' without the knowledge or consent of the Cabinet.

Mr Nicolson's novel is a masterpiece of subconscious (or is it perhaps deliberate?) 'low-down' which shows how a crisis can arise owing to laziness, apathy, competitive preoccupations and lack of cohesion amongst responsible Ministers. In war-time the semblance of a united Cabinet is thought even more vital than in peace-time, but this does not reduce the possibility of permanent truthfulness in Mr Nicolson's entertaining picture. We may perhaps repeat our initial question: How far does the *whole* Cabinet endorse the obliteration bombing policy of Sir Arthur Harris?

5 Future Prospects

It seems certain that, unless our leaders or we ourselves call a halt to this policy, it will be carried through to its terrible end – whatever that may be. In the *Daily Herald* for November 5th, 1943, an article by Wing Commander Charles Bray gave some indication:

> The air forces of the Allies are now ready to strike with their heaviest and most continuous attacks at Germany, and the morale of the German people and the blows will fall from many directions. Already Germany has had the first taste of what is to come. Here is the evidence. In the 24 hours ending yesterday morning, over 4,000 tons of bombs were dropped on Germany and occupied Europe. Over 1,500 four-engined bombers were probably over German territory during that period.

He added, under the heading 'This Is Still To Come':

> Three-pronged pincer bombardment of every industrial centre of the Reich and the occupied countries from Britain, Russia and Italy is considered to be a natural and immediate sequel to the Moscow Conference. Shuttle assaults with British, Italian, and Russian airfields as termini and turning points may be expected.

The following day, another characteristic pronouncement was made by Sir Arthur Harris himself during a speech in Northamptonshire:

> We propose entirely to emasculate every centre of enemy production, 40 of which are centres vital to his war effort and 50 that can be termed considerably important. We are well on the way to their destruction.

These 90 centres, said Sir Arthur, were all in Germany. Others in Italy and occupied territory would be 'treated separately'. To carry out the experiment of bringing Germany to her knees by bombing alone, we are thus committed to reduce to ruins 90 great cities, with their museums, libraries, hospitals, colleges, schools, churches – and human beings. That this can be done, *The Spectator* of July 30th, 1943, left us in no doubt: 'Thanks to the vast American production, the scale [of air attacks] can still rise. It is over twice what it was a year ago; a year hence, if the war still requires it, it will be twice as much again.'

That it is well on its way to being done, we were assured by Basil Cardew, *Daily Express* Air Reporter, on January 3rd, 1944:

> Today 24 of the essential towns have been more or less knocked out. At least 20 others have been severely damaged. Roughly in these ten months, German cities have been disappearing at the rate of two or three a month.

We who were accustomed to state, during the 20 years' truce, that another war might mean the end of civilisation, seem likely to have been right. Let us now consider how far human life and treasure have already been destroyed by our raids.

Notes

1 According to Air Commodore Howard-Williams (*Daily Telegraph*, September 20th, 1943), each of these bombs:

> lay flat an area of about 70 yards and are effective against even steel and concrete buildings. These bombs go off on impact on the top of the buildings they hit and blast their way downwards as well as sideways.

Presumably the writer means 70 yards square.

2 On March 5th the use of a new 12,000lb bomb by the R.A.F. was announced in the Press. So far, according to the published statements, this has been used only for bombing factories.

3 This unBritish habit of threatening appears to have been adopted from the Nazis by the United Nations and is now used by our Allies as well as ourselves. For instance, General Arnold, commanding the U.S. Air Forces, was reported in the *Daily Herald* of December 13th, 1943, as saying:

> The European and Mediterranean theatres are getting bombers in numbers we did not dream of last year. Every town and village in Germany will be hit. It will be a fearful, terrible winter for Germany. And next spring we will double the intensity.

3

The Bombing of Germany

At the end of October 1943, according to the Berlin correspondent of the Stockholm paper *Afton Tidningen*, the German Ministry of Home Security disclosed that 102,486 persons were killed in R.A.F. raids on 12 German towns in the 7 months from April 1st to October 25th, 1943:

> The towns were: Hamburg, 28,350; Cologne, 18,146; Dortmund, 15,008; Hanover, 6,320; Düsseldorf, 6,205; Bochum, 4,829; Duisberg, 4,763; Wuppertal, 4,635; Mannheim, 4,368; Nuremberg, 3,947; Frankfurt, 3,184; Kassel, 2,731.
>
> *These figures do not include bodies which could not be identified.*
>
> According to a member of the German Government Statistics Office in Berlin, 1,200,086 German civilians were killed or reported missing believed killed in air raids from the beginning of the War up to October 1st, says a Zurich message. The number of people bombed out and evacuated owing to air raid damage was 6,953,000 . . . (*News-Chronicle*, October 29th, 1943; italics mine)

Since the publication of these figures, the R.A.F. has 'obliterated' Berlin in a series of mass raids of which the first 13, according to the Swedish newspaper *Allehanda*, were resposible for 74,000 deaths (*Daily Telegraph*, January 31st, 1944). Other major cities such as Bremen, Frankfurt, Leipzig, and Magdeburg have also suffered a similar fate.

The number killed by German air raids on Britain from the beginning of the War up to October 31st, 1943, is just over 50,000.[1] Apart from all that we have done to Italy and to German-occupied countries, our reprisals mean that on Germany alone, up to the end of October 1943, we had already inflicted more than 24 times the amount of suffering that we had endured. No doubt there are many non-adult minds which find reason for satisfaction in the anguish that we have caused to the enemy. But others will reflect more responsibly that each one of those million dead (to say nothing of the injured and seven million homeless) have relatives and friends who will remember. Their memories will be even more dreadful than those of the post-war blockade in 1919, which was a chief origin of Nazism. We shall have to reckon with those memories when the days of rebuilding come.

On May 29th, 1943, the German wireless gave the following details of non-military buildings completely destroyed in air raids: churches, 133; schools, 191; hospitals, 108; while 494 churches, 920 schools, and 231 hospitals were 'heavily damaged'. The number must have risen enormously since then, for in his speech at Cheltenham on November 5th, 1943, Sir Archibald Sinclair disclosed that during May, June and July 1943, Bomber Command dropped over 52,000 tons of bombs. He added that while 5 per cent of Coventry was destroyed in the German attacks of 1940, 40 per cent of Essen had been virtually destroyed, 54 per cent of Cologne, and 74 per cent of Hamburg (*The Times*, November 6th, 1943). Later, in the House of Commons on January 19th, 1944, Sir Archibald stated that in 1943 Bomber Command dropped over 136,000 tons of bombs on Germany, compared with 2,400 tons dropped on this country by the Nazis.

According to the *News-Chronicle* of July 3rd, 1943, the total bomb tonnage dropped by the Luftwaffe on British cities in the peak year of 1940–41 was approximately 35,000 tons. At no period did the Blitz reach an average of 750 tons a night. The biggest raid was about 450 tons on London on one night of 1941. In April 1941, one of the most intense months of raiding on Britain, the Nazis dropped about 6,000 tons (i.e. less than half of the R.A.F. bomb-load in June 1943).

In the second half of October 1943, an article by Air Commodore Howard-Williams, Air Correspondent of the *Daily Telegraph*, revealed that in the 100 days and nights from July 9th to October 17th no less than 74,000 tons of bombs were dropped on Germany and German-occupied Europe by the R.A.F. and the U.S. Army Air Force. Of this total, Bomber Command dropped 56,000 tons in night attacks, 48,000 tons being dropped on targets in Germany. The writer justifies these attacks by quoting an R.A.F. commentator to the effect that 'the enemy's industrial cities were now great labour camps in which the houses of the workers were virtual barracks'. The commentator did not mention how many children lived in these 'barracks'.

Later, Air Commodore Howard-Williams informed us that

> half of Germany's principal cities have already been heavily bombed. Some 17 of them have been very severely mauled indeed . . . For

> instance, Hamburg has had the equivalent of at least 60 'Coventries', Cologne 17, Düsseldorf 12, and Essen 10.

On December 8th, 1943, addressing the Royal United Service Institution, Air Vice-Marshal Saundby, Deputy Chief of Bomber Command, said that in the German cities so far attacked – and few except those in the far east and south-east had escaped – nearly 25 per cent of the total built-up area had been devastated. The number of buildings destroyed in Germany ran into millions. In the ruins of Hamburg, Düsseldorf, Cologne, and other great centres of production, civilized life as we knew it was no longer possible. Many of our recent attacks on Germany had reached a rate of 120 tons per square mile per hour, or 80 times the intensity of the heaviest raid on London. Many industrial towns, such as Barmen, Elberfeld, Crefeld, Remsheid, Aachen, and others, had been virtually destroyed with all their factories and gasworks and *everything else* in a single concentrated attack (*News-Chronicle* December 9th, 1943. Italics mine). 'Everything else' presumably includes the population – to say nothing of the wealth of historic treasures in such ancient towns as Aachen.

Eighteen months ago, in the *Sunday Express* of September 13th, 1942, Mr John Gordon assured us that 'for the first time the Germans are really beginning to squeal'. They have not yet 'squealed' to an extent that terminates the indescribable ordeal of their mothers and children. How much longer will the British people consent to this infliction, in their name, of wholesale massacres which even their leaders regard as experimental?[2]

'If the growing horror of air war, which surpasses all powers of imagination, does not diminish, the day must come when the limits of endurance are passed' (*Sunday Express*, August 29th, 1943). Are we really willing to wait and watch without protest until those limits of endurance are reached? As I have already pointed out, not anger and revolt, but apathy, fatigue, and a mechanical endeavour to save what can still be saved, are the immediate results of concentrated bombing. The breaking-point – difficult in any case to reach with a Gestapo-ruled people – is still far away.

Hamburg, we are told, has had the equivalent of at least 60 'Coventries'. Those of us who went to Coventry shortly after the raid of November 14th, 1940, know what that cruel attack meant to a once lovely historic city and its inhabitants. Visitors to cinemas who have seen the American film on the Battle of Britain will remember the vivid reproduction of the Coventry raid, the long procession of weeping mourners to the cemetery, the children's coffins, the gravestone marked 'Mother'. Do the inhabitants of Coventry really enjoy the thought that the citizens of Hamburg – the most anti-Nazi city in the Reich, with its once large Jewish population – have suffered 60 times as much as they did? Does it really fill them with glee to reflect that 60 times their number of young children, expectant mothers, women in childbirth, invalids, and aged people have perished in terror and anguish?

Some of them have already made clear that it does not. On November 30th, 1940, the following letter appeared in the *New Statesman*:

> Sir, Many citizens of Coventry who have endured the full horror of an intense aerial bombardment would wish to dispute statements made in the *Daily Express* to the effect that all the people of Coventry expressed the opinion that they wished to bomb, and bomb harder, the peoples of Germany.
>
> This is certainly not the view of *all* or even the majority of the people of Coventry. The general feeling is, we think, that of horror, and a desire that no other peoples shall suffer as they have done. Our impression is that most people feel the hopelessness of bombing the working classes of Germany and very little satisfaction is attained by hearing that Hamburg is suffering in the same way as Coventry has suffered.
>
> Margaret M. Evans, Arthur Jones, Evelyn J. A. Evans, J. D. Dugund, N. M. Caine, J. R. Sidgwick. 'Hollycroft', Fife Road, Coventry.

I refuse to believe – and indeed I do not believe – that the kindly people of Britain have changed profoundly in three years. What I do believe (as the *New Statesman* of December 18th, 1943, suggested) is that they either do not know the facts, or, where they suspect the truth, they have consciously put shutters over the windows of their imagination. Many deliberately turn their backs upon knowledge, ashamed and fearful of accepting the realities which a determined facing of the facts would disclose.

I now propose to give some of these facts, as revealed in the Press of our own and neutral countries.

Notes

1 The exact figure is 50,088 killed, this total being made up as follows:

22,842 men; 20,288 women; 6,424 children; 446 'unclassified' (presumably includes missing, believed killed).

2 They [Harris and Eaker] have assured their military superiors that Germany can be bombed out of the War this year. . . . Winston Churchill stated the reaction of the global strategists when he said: '*The experiment is well worth trying so long as other measures are not excluded.*' (*Time*, June 7th, 1943; italics the editor's)

4
The Facts Behind the Figures

What actually happens when 1,500 tons of bombs hit a town? A 'Special Correspondent with exceptional knowledge of the state of Germany's bombed towns' endeavoured to answer this question in a *Sunday Express* article on November 7th, 1943:

> Everyone knows the obvious results, as the reconnaissance camera shows them – the great cloud of smoke and then, when the air is clear again, the miles of roofless or ruined buildings.
>
> Viewed from such a height, the scene merely suggests a dead city, but in fact we know it must be a scene of frenzied activity, like that of an anthill which has been kicked open.
>
> Our own experience in the Blitz of 1940–41 gives only a little help towards imagining the situation over there in Germany.

The disproportion, he continues, between blitzes on Britain and mass air raids on Germany is too great for accurate comparison. But a German official report on the state of a German city after a great raid might run something like this:

> Owing to the destruction of water mains and the exhaustion of static water in all parts of the town, the situation was still critical three days after the attack, and several fires were still out of control. Much hosing and fire equipment were destroyed, and skilled workers were needed for repairs.
>
> There were heavy casualties among the personnel of the fire brigade, which is now much under establishment. Reinforcements from other areas were on this occasion quite inadequate. Billeting was impossible within the town, and all arrangements for feeding the reinforcements broke down – demolition, rescue work, etc. is therefore proceeding very slowly . . .
>
> Because of contamination by sewers and gas, all water for domestic consumption must be boiled. As the result of 900 fractures and major damage in the town gas works, the gas supply is entirely disorganized; most domestic consumers are unable to boil water and hot meals cannot be provided.

> Telephone communications failed throughout the area ... Despatch riders were used for all but the most urgent communications, but because of dislocation of road traffic, etc. it was later decided to set up local headquarters outside the town area. This led to a severe fall in public morale, and the suggestion spread that a second air attack would soon be made because the headquarters staff had evacuated to a safe area.
>
> Twelve hundred retail shops being out of commission, food distribution was gravely hampered. Disorganization of supplies affected the food position more than actual destruction of stores, though this also was serious.
>
> Lowered morale mostly takes the form of apathy; the population appears dazed and stunned.

It is noteworthy that this imaginary but factually based report avoids any detailed reference – which might disturb even the hardened readers of the *Sunday Express* – to the sufferings of the injured and the disposal of the dead. The article concludes by telling us that there are some 30 large towns in Germany, in many of which these conditions have already occurred, and that 'we really do not know, and can hardly imagine, what life in Germany can now be like'.

In a number of these towns, through the reports of neutral correspondents, foreign observers, repatriated prisoners, and R.A.F. personnel, we have at least a limited idea of the actual facts. I shall now, therefore, examine the effects of our raids upon 14 sample German cities or areas, dealing with them in alphabetical order for convenience of reference.

Some Sample Cities

1 *Aachen*

Aachen (Aix-la-Chapelle) was a relatively small and pleasant industrial town of 160,000 inhabitants, situated on the Belgo-German border. Besides being industrial it was also full of historical monuments, one of the most famous being the Basilica with its relics of Charlemagne, who made Aachen his capital and was buried there in 814.

The town has been attacked by the R.A.F. several times, the Basilica being reported damaged after a raid during the summer of 1941. How badly it has suffered since I do not know, but the German radio has announced that on the night of December 23rd, 1943, and again about a fortnight later, the *Dom* was hit and damaged. In one heavy raid (July 1943), several 8,000lb bombs were dropped, and a few hours later smoke from the fires had risen four miles high. Berlin radio reported: 'The streets of Aachen are burning. Flames leap up from every house. The streets are full of rubble, splintered glass and burning beams.'

Long before this raid, Sir Archibald Sinclair, quoted in *The Times* of March 5th, 1942, had said: 'Aachen and Münster are certainly in worse condition than Coventry and Plymouth.' According to *The Times* of January 1st, 1943, 30 per cent of Aachen (i.e. nearly 160 acres, which is larger than the devastated area of the City of London) had already been destroyed at that time.

2 *Berlin*

Even before the major onslaught which began on November 18th, 1943, Berlin had suffered over 90 raids. The one hundredth attack took place on January 2nd, 1944. Like all *large* bombed cities, it has displayed an astonishing power of recovery.

After a great R.A.F. raid on March 1st, 1943, 1,000 people were reported killed, fires were still burning three days later, and the Friedrichstrasse had 45 craters. Reporting this attack, the *Daily Herald* (March 20th, 1943) reminded its readers that only twice during the raids on London – on April 16th and 19th, 1941 – were over 1,000 deaths reported.

On June 10th, 1943, after further raids, very full details of damage throughout the city were published in the *Daily Telegraph*. Its correspondent wrote: 'As one of my informants put it with eloquent brevity, "Berlin's West End looks more like a battlefield than a city".'

On August 11th, 1943, after the heavy raids on Hamburg, the same newspaper reported that thousands in Berlin were

> unable to sleep for fear that sirens will announce the grand attack . . . They conjure up the horror of concentrated raids, in which thousands die of hunger and thirst when buried unhurt beneath the masses of rubble and masonry covering cellar exits.

On August 23rd, a very heavy raid did occur, in which 1,700 tons of bombs were dropped in 50 minutes. On August 25th, 26th and 27th, paragraphs in the *Daily Telegraph* described the after-effects. 'From Leipzigerstrasse to the Chauseestrasse looks like No Man's Land,' reported its Stockholm correspondent. One traveller stated:

> I have lived many years in Berlin, yet at no time during that drive [in a taxi through the city] could I identify which street we were passing through. There were just ruins, shattered all, and fire wherever we passed . . . From Friedrichstrasse, down Belle Alliance Platz, the whole of the Tempelhof district had been reduced to a wilderness.

According to reports from Berne, the first official Berlin police estimates of the casualties in this raid put the dead at 5,680.

Further heavy raids occurred on August 31st and on September 3rd, when 1,000 tons were dropped in 20 minutes. The *Daily Telegraph* of September 20th, 1943, carried a long description of the consequences of these attacks, from which the following extracts are taken:

> A picture of Berlin as it is today is given by a Swiss eye-witness article in the *St Gallen Tagblatt*, in which he states: 'The last air attack on Berlin inflicted, particularly in the West End of the city, colossal damage, and also in the inner city and at the southern end at Lankwitz and Lichterfelde. In these districts streets were hardly negotiable . . .
>
> 'Efforts have been made to save people buried under the debris by tunnelling from the neighbouring houses, but if this is too difficult nothing is undertaken as it is assumed the imprisoned people are dead owing to burst gas and water pipes.
>
> 'It was nerve-shattering to see women, demented after the raids, crying continuously for their lost children, or wandering speechless through the streets with dead babies in their arms.'

These devastating raids, however, were themselves mere curtain-raisers for the 'grand attack' which began on November 18th, when a force described by the *Daily Herald* as 'the greatest number of four-engined bombers ever to raid Germany' dropped more than 2,000 tons of bombs on Berlin and Ludwigshafen. The evening papers next day carried the headline: '350 Block-busters Flung on Berlin'.

This onslaught was followed on November 22nd and 23rd by further huge raids which turned Berlin into 'the most bombed city in the world'. In three raids together, 5,000 tons of incendiaries and high explosive were rained on the city. The *Daily Herald* of November 24th announced the second great raid under a big photograph of four grinning pilots, which struck the note of jubilation that was this time indulged in by the entire large-circulation Press.

The *Daily Telegraph*, as usual, carried the most comprehensive details of the raids, and gave the fullest description of their meaning for the tormented civilian population. Describing the attack of November 22nd as 'very nearly the heaviest raid on any target in the history of air warfare', the newspaper continued:

> Reports from neutral capitals last night made it clear that the havoc was on an unprecedented scale, particularly in the centre of the city . . . Thousands were reported killed and injured . . . Unbroken heavy cloud lay along the whole route . . . We bombed Berlin 'blind'. The bombers followed the brightly lit 'target indicators' . . . although the target area itself was not seen.

On November 25th, after the third great raid, the same newspaper continued the story by quoting a Swedish business man, the first air passenger to arrive in Stockholm from Berlin after living through the onslaughts of November 22nd and 23rd. He reported Berlin as being 'ten times worse today than it was yesterday. The Berlin we know has simply ceased to exist.' Ossian Goulding, the *Daily Telegraph*'s Special Correspondent at Stockholm, described the 'red-rimmed eyes and white, lined face' of the speaker. Continuing to quote him, this correspondent wrote:

> 'The fire brigades and A.R.P. personnel are powerless to cope with the situation. Day has been turned to night by the billowing clouds of evil-smelling smoke which fill the streets . . . *Unter den Linden* is a shambles today, there are long lines of burning buildings in it . . . The University State Library is still burning . . .
>
> 'Block-busters freed a number of wild animals from Berlin's zoo. Troops turned out with rifles and machine guns to hunt leopards, elephants, bears, tigers and lions in the Tiergarten . . . Men who should know estimate that 85 per cent of the suburb of Spandau has been wrecked. The situation there is so serious that it has been decided to evacuate the whole district . . . I saw wretched creatures, trapped by flames, hurl themselves from fourth-floor windows to their deaths. Asphalt in the streets is alight everywhere, while over all lies the stench of phosphorus bombs.
>
> 'I would describe the morale of the city as fatalistic, exhausted and grim, yet determined to stick it. I saw no panic and no demonstration of any kind, although I heard several persons become hysterical in the shelter where I took refuge during the actual raids.
>
> 'I cannot agree either that there has been any voluntary mass evacuation of the city. Certainly thousands of people have moved to

> the suburbs or to the country beyond, but more to find a roof over their heads than because they were actuated by fear . . .'[1]
>
> 'As for myself, my only concern now is to sleep round the clock. I never knew a possible air attack could be like this: that it could affect nerves, digestion, eyesight, and everything in this way.'

Another *Daily Telegraph* correspondent from 'Somewhere in Europe' reported:

> The destruction . . . is almost impossible to describe. Whole streets are ablaze. The heat is so fierce that people are collapsing because of it . . . Tens of thousands of people are leaving the city . . . Their faces are blackened with soot and smoke. Many of them have bandaged hands, signs that they were burned in frantic and useless efforts to put out the flames of the thousands of fires that raged last night and the night before . . . In the burning areas people can be seen vainly trying to save what is left of their belongings . . . The three raids coming on top of one another have stunned the people. Nazi propaganda that the people of Berlin cursed the R.A.F. is wrong. Instead, after the raid, the people of Berlin could find little to say. They only picked up what they had managed to gather and moved silently on . . .
>
> There are various figures of casualties. One says that 25,000 were killed on Monday night and the same number on Tuesday night. The Swiss newspaper *Die Tat* states that between 20,000 and 30,000 bodies, victims of Monday night's raid, have already been recovered . . .
>
> The population is still so stunned that it is too early to gauge their reactions.

On November 26th the *Daily Telegraph* reported Sir Arthur Harris as saying that the bombing of Berlin would continue 'until the heart of Nazi Germany ceases to beat'. That night the third heavy raid of the week, and the fourth in eight days, was made on the German capital. According to Denis Weaver, Stockholm Correspondent of the *News-Chronicle* (November 27th), one Swedish woman, Miss Rosberg, who was employed by Baron von Essen, was so tired by this time that after the 'All Clear' she lay down 'like thousands of others', and slept in the street amid the wreckage. Other observers reported that 'phosphorus rained from the sky. Many victims' clothes were burnt into rags.'

On the same day, the *Daily Telegraph* reported:

> Hospitals are crammed full with 5,000 or 6,000 hovering between life and death, and the medical staffs hopelessly overworked almost

> entirely with healing and treating burns . . . In the streets at night you see the homeless sleeping in wrapped blankets among the ruin of shattered homes.

At the end of November, a grave shortage of food and water was the subject of emergency conferences in Potsdam. 'Berlin is desperately short of water,' stated the Stockholm correspondent of the *Daily Mail* on November 29th. 'Many Berliners can obtain supplies only by queuing up at hydrants where the mains are still intact.' A fortnight later, the *News-Chronicle* (December 5th) recorded that 500,000 Berliners were suffering from 'cellar sickness' – a form of influenza caused by hours spent in damp cellars during raids. In the first week of December, 2,000 deaths occurred from this cause.

On December 20th, Olle Ollen, Berlin correspondent of the Stockholm paper *Social Democraten*, reported that Berlin was facing a 'spartan' Christmas even if no further heavy raids occurred. He stressed the fact that blitzed Germans now comprise a new 'army' of sympathizers with the Reich's cause. 'It may be', he said, 'they do not always sympathize with the Nazis, but they do with Germany.'

Between the end of November and January 20th, 1944, six further heavy raids occurred, making ten in two months. Between November 18th and January 4th, more than 14,000 tons of bombs dropped on Berlin. On Christmas Eve a 1,000-ton raid was launched at 4a.m., and the city – perhaps trusting to the hope of a truce during the Christmas season – was taken by surprise. The *Evening News* that day reported an experienced squadron commander as saying: 'We had them absolutely foxed.' The raiders, directed by Pathfinder squadrons, concentrated on areas known to be relatively untouched. Commenting on the day chosen for the raid, Paris radio said: 'There are no words to describe the feelings in the hearts of men at this new crime, which calls for a punishment.'

Quoting a cable from Robert Vivian, Reuter's Special Correspondent in Washington, the *Sunday Express* of December 26th, 1943, reported:

> Military circles in Washington believe that instructions have been issued to the R.A.F. and the American Air Forces based on Britain to complete the wiping out of Berlin before invasion starts from the West.[2]

A 2,000-ton raid followed on December 29th, the third anniversary of the fire raid on the City of London at the end of 1940. Yet Sir

Archibald Sinclair assured his Plymouth audience on January 22nd, 1944, that we do not go in for reprisals, and continued, with manifest self-contradiction, to state: 'It is too late now for them to pretend that the medicine we are administering to Germany is not their own.' What reason we have for adopting the 'medicine' – and the standards – that we ourselves have repeatedly condemned, he did not explain.

On January 4th, 1944, the *News-Chronicle* disclosed a calculation by Bomber Command that another 15 to 20 big attacks would be needed 'to finish the job'. The previous Sunday (January 2nd) the Air Correspondent of the *Observer* had remarked that 'It will probably not be necessary to devastate more than 12,000 of Berlin's vital 18,000 acres to end its wartime life as an organized city.' The article continued: 'Some of the districts known to have been bombed are among the most densely populated in Europe.'

After the raid of January 2nd, Denis Weaver reported great loss of life through the failure of alarm signals, as many sirens went down with the crumbling buildings from which they had been operating. He quoted the Gothenburg *Handelstidningen* as estimating that half a million people had lost their homes in the raids of December 24th and January 2nd. 'Overcrowding is now critical. As many as ten or more people share the same room. Health conditions are worsening' (*News-Chronicle*, January 5th, 1944).

The Times of January 11th, 1944, recorded that 17 per cent of all the buildings in Berlin were destroyed in the six major attacks between November 18th and December 16th:

> It is thought probable that neutral estimates that a quarter of the capital has been devastated may prove reasonably accurate. Since the photographs were obtained the R.A.F. have dropped a further 5,000 tons on Berlin.

On January 20th, in the tenth and largest raid up to that date, another 2,300 tons went down. According to the *Evening Standard* of January 25th, an unconfirmed German underground radio report said that there were now more than one and a half million bombed-out people in Berlin:

> The hospitals are crowded with wounded and the number of deaths is mounting daily [the report said]. The German capital's last raid was as heavy and devastating as the great attack on November 22nd . . . The Rummelsburg waterworks were so badly damaged that North Berlin is without water.

On the same date, the same page of the *Evening Standard* published a warning by Dr M. D. Mackenzie, a Ministry of Health Medical Officer, that after the War 'we have every reason to fear extensive outbreaks of typhus fever, malaria, and water-borne diseases such as cholera'. For this additional terror, Europe may well have Bomber Command to thank.

On January 27th–28th, 28th–29th, and 30th, three further major raids occurred, bringing the total of these mass attacks on Berlin up to 14. A traveller reaching Stockholm from Siemenstadt (South Berlin) was reported as follows by the *Daily Herald* on January 29th:

> I could not believe that the ruins I saw were those of the same town I had seen an hour before. It was so terrible that I have not been able to eat since.
>
> Worse than the damaged factories and houses was the stream of bombed-out people plodding wearily along the roads, and the rescue workers and ambulances carrying away the dead.
>
> Fires in the outer districts were started by explosives. The flames created currents, the draught from which blew through the streets, fanning smaller fires into bigger ones.

Next day, January 30th, the *Sunday Express* described Berlin as 'a Concert of Hell', and reported that early the previous morning the German capital went through 'one of the most terrible periods since the English raids of destruction . . . began'. *The Sunday Times*, quoting its Stockholm correspondent, said:

> Owing to lack of stretchers and first-aid personnel, the injured often had to lie for several hours on the pavements in the cold and the rain.
>
> Hospitals are overcrowded and cannot take in more casualties. Temporary hospitals have been established in two large factories, but doctors, nurses, and medicine are lacking.

After the raid of January 30th, the *Evening Standard* on February 1st reported the Scandinavian Telegraph Bureau as saying: 'The last 48 hours have been far worse than all the November raids. The giant city of Berlin is dying, slowly and painfully, amid scenes of indescribable destruction.'

The *Daily Herald* added the following day:

> Berlin is a city of sleepwalkers. People, dazed by the succession of blitzes, are wandering aimlessly about the streets. They have nowhere

> to go, nothing in which to ride, nothing to see except ruin, death and decay . . . Berlin, heart of the Reich, has almost stopped.

At the beginning of the Berlin raids, on November 28th, 1943, Frederick Tomlinson, Air Correspondent of the *Observer*, commented as follows:

> The scientists who have developed our newest bombs and our latest aircraft equipment have presented us with a terrible weapon, the logical purpose of which is not so much to destroy industrial buildings as isolated objectives but *to make industrial life* with its attendant war production *impossible in all the large cities of Germany*.
>
> Bomber Command is confident that if it is adequately supplied it can achieve its object, though it may require not the 13,000 tons (which have already been dropped on it) but 50,000 tons to destroy a target such as Berlin.
>
> *The military value of bombing by night will be debated to the end of the War*. (Italics mine)

Not debatable, however, are the death, injury, and suffering of huge populations, and the psychological aftermath which the memory of these things will create in Europe.

3 *Brunswick*

As an industrial town, Brunswick was comparable with Coventry. In 1939 their respective populations were: Coventry, 220,000; Brunswick 201,306. But, like Coventry itself, Brunswick had a value quite other than that conferred by its industries. Its *Dom* (1172) contained the tomb of Henry the Lion, Duke of Saxony, and its ducal palace possessed collections of pictures, majolica and gems which were among the finest in Germany. Spohr, the musician, was born in the town, and Lessing died there (1781).

On January 15th, 1944, according to the *Sunday Express*, this relatively small city received in 23 minutes 'the most furious air assault of the War . . . It worked up from an average of 90 tons a minute to the unprecedented weight of 150 tons a minute.' The account of the raid, occupying the main space on the front page, received huge headlines: 'The Most Terrible 3 Minutes' Bombing Ever Known – 450 tons'; 'More Than London Had In Any One Night'; '2,000 Tons Altogether in Brunswick "Saturation"'. The article continued:

> Never before has a town been so saturated . . . During the peak of the attack a great force of Lancasters in three minutes dropped a heavier load than the Luftwaffe have ever dropped on London in a whole night . . . The airmen brought back accounts of colossal destruction. The target was cloud-covered, but the area to be destroyed was clearly ringed with sky-markers . . . Goering's night-fighters did not arrive over the target until the attack was over. They had been sidetracked by a feint attack by Mosquitoes on Berlin and on Magdeburg.

Reporting this attack further on January 17th, the *Daily Herald* stated that 'many direct hits' had been made on the big Messerschmitt works, which were presumably the excuse for destroying the town. It did not state that these works had been destroyed, or even effectively put out of action. The report however added:

> Of the city's 250,000 inhabitants, 50,000 are said to be homeless. The death roll is estimated at 12,000, but raid casualty reports allowed to leak out of Germany have frequently proved grossly exaggerated.
>
> Several barracks for foreign workers are reported to have been destroyed, and the 'slave' workers themselves to have fled to the Harz Mountains in spite of the snow and bitter cold.
>
> Brunswick, according to strictly censored reports, has practically ceased to exist.

We may well ask how many conscripted workers from friendly countries perished with the town.

4 *Cologne*

Cologne, a city of 750,000 inhabitants, was one of the most beautiful and historical towns in Germany. 'The glory of Cologne', says Nelson's Encyclopædia, 'is its Cathedral, one of the noblest and most impressive examples of Gothic architecture in existence. Its foundations were laid in 1248.' In the narrow, crooked streets in the centre of the historic area, there were many houses of the fifteenth and sixteenth centuries, and from even earlier.

As the chief city of the Rhineland, Cologne has suffered repeated raids which have given it the equivalent of '17 Coventries'. On May 30th, 1942, it was selected for the first of Sir Arthur Harris' 1,000-ton 'saturation raids'. *The Times* of January 1st, 1943, reported a senior officer of the R.A.F. as saying: 'The City of London contained less than 120 acres of devastation. In Bomber Command's big raid on Cologne, more than 600 acres were devastated.'

During the summer of 1943, the R.A.F. in three great raids destroyed more than 80 per cent of the central city area, and 75 per cent of the other fully built-up districts on the west bank of the Rhine. According to the *Evening Standard* of June 30th, German overseas radio said that in addition to damage to the cathedral, the celebrated Roman Church of St Cuthbert was also a victim of the bombs. It put the number of churches destroyed in Cologne at about 35.

On August 5th, the *Daily Telegraph* quoted a detailed report of the raid havoc, given to the Swiss newspaper *St Gallen Tagblatt* by a Swiss citizen who had just returned from Cologne, where he had lived since 1936. The inner city, he said, was finally destroyed on the night of June 28th when, it was estimated, nearly 1,000 tons of bombs were dropped. The previous month a 'complete job' had been made of the suburbs, when whole working-class areas were razed to the ground.

Speaking of the actual damage done during the raids, this witness said:

> Except for the Cathedral and a few isolated houses the old and inner city of Cologne has ceased to exist.
>
> Among buildings destroyed are the Town Hall, the Hansa Hall with its well-known Gothic façade, and the Wallraf Richarts Museum . . .
>
> Around the Cathedral, barely 30 yards off, all hotels and business premises have been burned to their foundations. The Savoy Hotel collapsed.
>
> From the big Buelheim suspension bridge I gazed on what ought to have been the panorama of Cologne. I saw only masses of thick smoke.
>
> As I rode towards the city I noticed that the trees along the Rhine were stripped of their foliage and covered with thick dust.
>
> I have always imagined that a prehistoric landscape without life must have been like this. The sight of human beings moving about in it gave me a cold shiver. I seemed to be on another planet.
>
> In Cologne itself people looked apathetic. They were too tired to talk. It was then only seven hours after the raid . . . In front of houses lay goods and chattels, and also people in a state of utter exhaustion.
>
> In the Glockengasse I came on a woman searching among a score of corpses for a relative. Further on in the city, in a big square, I saw bodies laid out in hundreds.

5 *Düsseldorf*

Once a bright, clean town with pleasant parks and friendly inhabitants, Düsseldorf was described by the *News-Chronicle* Air

Correspondent after the heavy raid of June 12th, 1943, as 'a dead city. It was killed in a night.' More than 380 acres of the town had, however, already been damaged by the end of 1942.

Thirty-six hours after the June raid, *The Sunday Times* commented: 'Raids on the scale of Friday night's attack on Düsseldorf mean the virtual blotting-out of the city as far as ordinary residential life is concerned.' On June 22nd, the *News-Chronicle* added: 'According to the well-informed Catholic Swiss newspaper *Vaterland*, 400,000 of Düsseldorf's population of 600,000 are homeless. Twenty thousand people have been killed.'

A neutral report by a Swiss correspondent in *Das Volksrecht* (Zurich) on October 2nd, 1943, ran as follows:

> Düsseldorf made the most frightful impression of all the western German cities. This once beautiful city is today a heap of ruins. The gaiety of its population has vanished. There are sad faces to be seen everywhere. The new railway station is completely destroyed. The station square with its great hotels and the main post office is covered with ruins. All the streets converging on the square show the same picture of destruction. The centre, north and south of the city have suffered most. All the entertainment buildings have disappeared: the Municipal Theatre, the Concert House, the Jaegerhof Castle, the Apollo Theatre, and all the great cinemas and department stores such as Tietz. The fashionable hotels such as Breitenbacher Hof, the Park Hotel, and also the Hochhaus, are completely burned out. A high police officer told me that 2,500 people were killed during one night of heavy bombardment and that the Provincial Fire Insurance building still covers its victims. 18,000 dwelling houses have been destroyed in Düsseldorf and 350,000 people rendered homeless. These figures do not include the destruction in the (industrial) suburbs of Gerresheim and Benrath.

Even under these conditions, however, the city was not dead enough for Bomber Command. On November 5th, 1943, an article by Wing Commander Charles Bray reported yet another 'obliteration raid':

> R.A.F. in great strength dropped over 2,000 tons of high explosive and incendiary bombs on Germany on Wednesday night. This brought the total weight of bombs dropped in the 24 hours to 4,000 tons, the greatest weight ever dropped in a day and night operation. Düsseldorf was the main target, and the raid was all over in 27 minutes. After the 2,000-ton raid on June 13th reconnaissance photographs showed that two-thirds of the central city had been

destroyed. It has now been raided 58 times. Nineteen of our bombers were lost.

6 *Frankfurt*

Frankfurt-am-Main was the birthplace of Goethe. In the older part of the city stood the house in which he was born, with a Goethe museum attached. Other historic monuments were the Römer, a collection of ancient buildings round a wide square, which served as the Town Hall and contained the hall in which the German Emperors were elected; the cathedral, founded in 850; the Saalhof; the Academy of Social Sciences, opened in 1901; the house in which Schopenhauer lived; and the ancestral home of the Rothschilds. On the opposite bank of the Main, the Städel Institute contained a picture gallery, art collections, and an art school.

Frankfurt, however, owed its special importance among German towns to its fame as a commercial and financial centre. It was, therefore, selected for a 2,000-ton 'saturation raid' as a Christmas present on December 20th, 1943. This, according to the *News-Chronicle* of December 22nd, 'may have been the city's death blow'. The raid lasted only for half an hour. *The Times* of December 22nd reported:

> At the height of the attack bombs were falling at the rate of 70 tons a minute, many 8,000lb and 4,000lb 'block-busters' among them. The raiders left a great oval of fires, stretching from east to west across the city, and pillars of smoke billowing up to a height of 14,000 feet.

Frankfurt had a population of over half a million. This, and subsequent 'saturation raids', have undoubtedly reduced that population by methods which compare with and may well exceed the terror and agony inflicted by poison gas.

7 *Hamburg*

Terrible as the facts already quoted are, the series of raids on Hamburg between July 24th and August 2nd, 1943, surpassed them all in horror. If we did not possess authenticated reports from neutral sources regarding their appalling consequences, we might well dismiss them as the evil nightmare of a homicidal maniac. The destruction of Hamburg, like the later mass attacks on Berlin, may testify to our capacity to win the War, but it also provides irrefutable evidence of the moral and spiritual abyss into which Britain and her rulers have descended during the past four years.

In eight heavy raids during ten days and nights, a total of about 10,000 tons of high explosives and incendiaries was dropped on this city of 1,800,000 inhabitants, completely destroying nine square miles (77 per cent) of the built-up area (*Daily Telegraph*, September 20th, 1943). 'Hardly anyone, it is alleged, escaped in the heavily populated area of many miles on which the Allies planted a carpet of hundreds of thousands of explosives and incendiaries . . . At least 20,000 perished in shelters alone' (*Daily Mail*, October 9th, 1943). One R.A.F commentator admits that 'the greatest destruction from these raids has been to business and residential property, especially in the built-up area' (*Daily Telegraph*, September 20th, 1943).

Photographs taken after these raids revealed that 'of Hamburg's fully built-up 4,000 acres, 1,700 have been destroyed, while of 3,400 less densely built-up acres, 1,900 have also been completely shattered' (*Daily Telegraph*, August 6th, 1943). An officer of the R.A.F. who had been over Hamburg said:

> The term 'raid' is no longer expressive enough for what is happening. From what I have seen in two of the six air attacks made within 72 hours the destruction is truly devastating. In comparison the enemy raids on London were child's play. What is going on at Hamburg can be repeated on any target we select. Hamburg is the first to be dealt with . . . (*Daily Telegraph*, August 29th, 1943)

Another R.A.F. commentator, describing the raids as 'the most striking bombing event in history,' said: 'To all intents and purposes a city of 1,800,000 inhabitants lies in absolute ruins . . . It is probably the most complete blotting-out of a city that ever happened.'

An eye-witness of the raids, writing in the Swiss newspaper *National-Zeitung*, reported:

> We passed whole streets, squares, and even districts . . . that had been razed. Everywhere were charred corpses, and injured people had been left unattended. We will remember those Hamburg streets as long as we live. Charred adult corpses had shrunk to the size of children. Women were wandering about half-crazy. *That night, the largest workers' district of the city was wiped out*. (Quoted in *Reynold's News*, August 8th, 1943; italics mine)

'A preliminary estimate of the killed in the eight heavy raids on Hamburg was more than 58,000' (A.P. message from Stockholm, *Daily Telegraph*, August 9th, 1943). The official German estimate, as is the habit with all official estimates (which are provided for home as well as for foreign consumption), put the figure much lower, though it admittedly left out the unidentifiable dead. Facts which I am about to record suggest that these must have reached exceptional proportions in Hamburg. A Swedish A.R.P. official, quoted by the *Daily Telegraph* on November 27th, 1943 (by which time it might be expected that the number of Hamburg casualties would be established), stated that 'nearly 50,000 were killed outright'. Eighteen thousand people were also reported to have been drowned when the Elbe Tunnel received a direct hit (Reuter message from Zurich, *Daily Telegraph*, August 25th, 1943).

A Danish consular official, interviewed by the Stockholm newspaper *Aftonbladet* after he had escaped from 'the most blitzed city in the world', said:

> Hamburg has ceased to exist. I can only tell what I saw with my own eyes – district after district razed to the ground. When you drive through Hamburg you drive through corpses. They are all over the streets, and even in tree-tops.

Swedish seamen who arrived in Malmoe alleged that not more than 50 houses remained standing in the whole of Hamburg. 'On Saturday,' they said, 'the destruction was so complete that not even the sirens were working' (*Sunday Express*, August 8th, 1943). Exaggerated as the first part of this statement may appear, a paragraph in the *Evening Standard* of November 29th, 1943, reported that 'It is now possible to drive for half an hour through the centre of Hamburg without passing a single house.' According to this information, which reached Reuter from Stockholm, 'A large part of the town was said to have been so completely devastated that "there is no point in clearing it".'

Other Swedish refugees described the terrible characteristics of our phosphorous incendiary bombs:

> They talked of the strange sensation of seeing gardens on fire in a city ravaged by flames. Hundreds of people were found burned to death in the streets and the clothing was scorched off many by the fires. About 47,000 dead bodies were counted before search work began, and estimates of people killed vary from 65,000 to 200,000 . . .
>
> The town is almost deserted. For a fortnight the fires have raged unchecked and the people are almost poisoned by the smoke and the ghastly smell that hang over the empty streets, where the walls are still radiating heat.

> Some of the refugees, who were wearing strange evacuee clothes like beach pyjamas, described the city as 'sheer hell'. Panic often broke out in large seven- to eight-storeyed shelters, which held 3,000 to 6,000 people during the terrific explosions. (*Daily Telegraph*, August 14th, 1943)

A stoker who deserted from a German ship corroborated these facts in an interview with the Stockholm correspondent of the *Daily Telegraph*. 'People', he said, 'went mad in the shelters. They screamed and threw themselves, biting and clawing the doors which were locked against them by the wardens ...' (*Daily Telegraph*, August 25th, 1943). Other reports stated that owing to the great heat of the fires, people died from suffocation in shelters. On September 20th, this form of torture was described in the following (translated) article by the Swiss correspondent of the *Basler Nachrichten* (*Basle News*). This article related how, as the result of a mass fire during one of the Hamburg raids, many more persons perished in a few hours than the total of air-raid victims in London since the beginning of the War.

When the article was first published in this country, its alleged scientific explanation of the phenomena which it described was questioned in the correspondence columns of the *Manchester Guardian* by a Manchester physicist, despite the fact that the article was published by Dr Oeri, a Member of the Swiss Parliament and a man of outstanding integrity. What I am here concerned with, however, is not the scientific explanation of the facts, but the facts themselves. The *Basler Nachrichten* makes clear, first that thousands of people in Hamburg died a terrible death which struck more ruthlessly at children and their mothers than at men; and secondly, that the nature of this mass slaughter was well known in Switzerland – with what effect upon the reputation of Britain as a civilized country, the editor judiciously avoids recording. That the suffocation and roasting alive of people in the shelters occurred in addition to the burning to death of others in the streets appears to be corroborated by further evidence given in this book – for example, the German report of the Hamburg raids quoted by the *Daily Telegraph* on August 12th, 1943 (*Seed of Chaos*, page 108), and the *Bruesseler Zeitung* account of the Allied raid on Hanover quoted by the *News-Chronicle* on November 11th, 1943 (*Seed of Chaos*, page 139).

What Happened in Hamburg
Report by a Swiss Correspondent.[3]

During the bombing of Hamburg there was a catastrophe in one densely populated part of the town of several square kilometres which eclipsed all previous happenings of the bombing war. It occurred as a result of the area being covered with mines, high explosives and phosphorous bombs and hundreds of thousands of ordinary incendiaries.

It must be emphasized that the effect was one which can only be achieved when bombing densely populated residential districts, but not when bombing factory districts. Every physicist of the air war could have calculated this effect in advance if the number of H.E. and incendiary bombs to be dropped on a given area were known to him. It is a question of the well-known fact that every open fire sucks in the oxygen it needs from the surrounding atmosphere, and that large fires, unless there is a strong wind, will lead to the creation of so-called air chimneys up which the flames will rush with ever-increasing force. If the area of the fire covers several square kilometres, then the flames licking out of individual rows and blocks of houses will combine into one big blanket of fire, covering the entire area and rushing up to ever greater heights. According to English reports, the Hamburg fire reached a height of six kilometres, that is, up to that height the heat rose in one compact body.

Under these conditions the following occurs: within the area of the fire a rush of air is created, reaching the strength of a typhoon. The effect is that of enormous bellows pumping air into this district from all directions; for the sea of flames sucks in air from its surroundings. In this, the streets serve as channels through which the air passes towards the centre, and at the same time the air rushing through the streets sucks the flames from the burning houses horizontally into the streets. Thus, human beings and flames will compete for the available oxygen and, naturally, a fire of this size will get the better of it. The flames suck the last remains of oxygen from all rooms, shelters and cellars, and at the same time devour all the oxygen in the streets.

The immediate result in the cellars is a shortage of oxygen and breathing difficulties for the people present. At the same time the temperature in the shelters rises unbearably, but the people are prevented from leaving the shelters during the early stages of the bombing by the constant rain of H.E., incendiary and phosphorous bombs, which release a fine shower consisting of a mixture of rubber and phosphorus. Experience has shown that when the people finally make up their minds to leave the cellars it is too late. They have no strength left to carry out their decision, and even if they have they lack the strength to resist the heat and the lack of oxygen in the street. It is easy to see that men, with their greater power of resistance and stouter clothing, are better able to resist such a method of attack than women and children. That is why the majority of the victims are women and children. Numerous completely charred bodies of women and children were found along the outer walls of the houses; women and children in light summer clothing who emerged from the cellars into the storm of fire in the street were soon converted into burning torches.

Naturally, hundreds and thousands of men too lost their lives in the streets of this district. Hamburg experts who are in charge of the salvaging of bodies have stated that only a minute percentage of the population residing there can have escaped with its life under the conditions prevailing during the attack. The whirlwind surrounded the entire district with a fiery wall and only those were able to save themselves who escaped at the very beginning. Even medium-sized squares and wide streets offer no protection.

The condition of the cellar shelters, which have meanwhile been opened, gives some indication of the temperature which must have prevailed in the streets. The people who remained in these rooms were not only suffocated and charred but reduced to ashes. In other words, these rooms which, without exception, became death-chambers for dozens and hundreds of people, must have reached a temperature such as is not reached in the burning chambers of a crematorium. One doctor who supervised the salvage of the bodies remarked that the incineration of the bones had in many cases been more complete in the cellars than it is in the normal process of cremation. Obviously, it is impossible to identify the bodies, as all the belongings of the people have also been reduced to ashes.

The 20,000 bodies salvaged so far in Hamburg come mainly from this district. Even today, the work of salvaging is still extremely difficult because the temperature in the cellars a fortnight after the fire is still such that any introduction of oxygen makes the fire flare up again.

The many reports by survivors of burning women and children, and of women throwing their children into canals, are, therefore, not invented. How great was the temperature prevailing in these streets is further proved by the fact that the glass in the windows and metal frames were reduced to ash and cinders.

As we have said, all this occurred in a strictly defined district of some kilometres square. Obviously, effects like those described can only be achieved in densely populated residential districts with high houses and relatively narrow streets. The streets, however, need not be

> very narrow, for roughly 50 women and children were found suffocated, half charred, and with all their clothing torn from their bodies by the storm, on a playing field which was situated at the centre of a street crossing. It appears, therefore, that the air war in this form can indeed turn entire districts of a large city, and, above all, the residential quarters of workers and employees, into a fiery grave which no one can escape who has not the courage to flee in the early stages through the rain of phosphorous, H.E. and incendiary bombs.

The extreme suffering in Hamburg – of which this article provides at least a substantial indication – has had psychological consequences which any intelligent student of human nature could easily have foreseen. During recent weeks, the German Social Democratic Party Executive in London has received a number of reports dealing with the moral and material effects of the R.A.F. offensive against Germany last summer and autumn. One direct report runs as follows:

> There was also criticism of the R.A.F. One of the functionaries referred to leaflets dropped over Hamburg in which various promises had been made. The reporter comments on that, saying that such language had been sorely missed before 1933, and it was not fair now to punish the defenceless civilians for what the allies themselves had fostered before the War. The reporter has lost everything in the raids. There was consternation about the almost complete destruction of Eimsbüttel and Barmbeck, *districts entirely populated by workers, with very few war factories.*
>
> *The reporters stressed their anti-Fascist attitude, but added that now, after the destruction of 'Red Hamburg', the 'City without Nazis', it was very difficult to point to the Western Powers as a hope for a better future.* (Italics mine)

The words 'Western Powers' are significant. How far, we may ask, are Sir Arthur Harris and Bomber Command playing straight into the hands of those who would gladly see Britain and the U.S.A. become the powers most hated in Europe after the War?

8 Hanover

Before the War, the old town section of Hanover possessed a number of fourteenth-, fifteenth- and seventeenth-century buildings, such as the former royal palace, the chancellery of justice, and the house of Leibniz, which had been converted into an industrial art museum. Its famous polytechnic, housed in the Welf Castle, was attended by nearly 2,000 students. Close by was the Herrenhausen Castle (1698), with its beautiful gardens, the favourite residence of Kings George I and II.

On January 3rd, 1944, the *Daily Express* Air Reporter stated: 'Hanover is two-thirds gone.' The following account had previously been received from Denis Weaver, *News-Chronicle* correspondent at Stockholm:

> Colourful details of the Allied raid on Hanover on October 8th and 9th appear in the *Bruesseler Zeitung* of November 3rd.
>
> The paper states that the attack was so intense, and the damage and fires so extensive, that police and air-raid wardens were forced to concentrate on the single task of getting people out of shelters which had become death-traps.
>
> Wardens ran into shelters and cellars telling shelterers that their only hope of life lay in getting out into the street and scattering.
>
> The paper adds the admission that many were too terrified to move and had to be taken out by force.
>
> Then began a series of dashes . . . to find a way out of the town through burning buildings, crashing masonry and falling sparks. Firemen had to prepare a path with hoses playing on the blaze.
>
> In some places people were taken to public tanks and told to dip their clothes in the water to make them less inflammable.
>
> 'The old city of Hanover', says the paper, 'is a pile of ruins in which none can now find his way.' (*News-Chronicle*, November 11th, 1943)

9 Leipzig

Reporting that 'nearly one-third of the most densely built-up areas' was devastated in the 1,500-ton attack by Bomber Command on the night of December 3rd, 1943, the *Daily Herald* of January 17th, 1944, described Leipzig as 'Germany's fifth largest industrial city'. It omitted to mention that this town, with its 800,000 population, is also the seat of a famous university, founded in 1409, and attended by several thousand students.

Like Brunswick, the city was taken by surprise owing to a feint attack on Berlin, and had to endure the raid without fighter protection. Quoting an account by a man who had just returned from Germany, Noel Panter, *Daily Telegraph* Special Correspondent, reported from Zurich on December 13th:

> The raid was terrifying. It lasted for 30 minutes which seemed a century.

> The Opera House, the Dresdener Bank, the main post office, and the Fair buildings were all in flames. All were destroyed . . .
>
> Without exaggeration one may say that the whole great mass of buildings which constituted Leipzig, city of 800,000 inhabitants, no longer exists.
>
> As a crowning misfortune all the Leipzig firemen had been summoned to Berlin to fight the fires there.
>
> At noon on the following day it was still dark in Leipzig, which was entirely enveloped in black smoke, and the air was almost unbreathable . . .
>
> Everywhere fires continued to rage and every moment the crash of collapsing buildings was heard. In the streets the inhabitants were fleeing terror-stricken under a rain of sparks. All had pieces of wet cloth or handkerchiefs wrapped round their heads to prevent their hair catching fire. They hurried to the large squares and open places to try to get some oxygen.
>
> The very large number of victims was due partly to the collapse of the shelters inside houses.

When the bombers reached Leipzig, reported the *Daily Herald* of January 17th:

> The whole of the city and the surrounding country was hidden by 10/10ths cloud . . . All that the crews could see was cloud. They bombed blind in the area marked out by the Pathfinders' target indicators.

By this means they achieved a 'great concentration of bombing in the most vulnerable part of Leipzig', in circumstances – now relatively familiar in R.A.F. raids – in which all pretence of attacking only or mainly military objectives had been abandoned.

10 Mainz

Mainz, in the grand-duchy of Hesse, was one of the most important commercial centres on the Rhine. It was also a city of unusual historic interest, having been founded, as 'Maguntiacum', in 13 B.C. Its importance began in 747, when it was made an Archbishopric. The picturesque cathedral in the Marktplatz dated in its more recent form from the twelfth to the fourteenth centuries, with later restorations.

In an R.A.F. raid carried out in August 1942, this cathedral, and many other cultural monuments, were burnt to the ground. According to the *Frankfurter Zeitung*, the Bishop's Palace was also seriously damaged, and five churches were obliterated (*The Times*, August 17th, 1942).

During November 1942, a visitor to the Exhibition at Messrs Rootes, in Piccadilly, of photographic enlargements showing the devastation caused by air raids on German and other continental cities, was given the following descriptions of Mainz by an R.A.F. official guide:

> The whole core of the city is destroyed. The total area of devastation within the city equals approximately 135 acres.
>
> This is the heart of Mainz, the largest area of concentrated devastation in any German city. Fifty-five acres in the city centre have been devastated. Ruined civic buildings include museums, churches, schools. There is hardly a house habitable or a building useful in the centre of the city. Shops, offices, art galleries are destroyed. Here a 4,000-pound bomb has fallen; what was a built-up area is now an empty space.
>
> This was the main street of Mainz. A whole line of offices in streets like Cheapside have been destroyed. The town theatre – the pride of Mainz – has been burnt out.

11 Münster

The old town of Münster, the capital of Westphalia, contained a beautiful Romanesque cathedral and many other architectural treasures. Of this city, a foreign observer was quoted as saying (*News-Chronicle*, May 17th, 1942):

> I have not seen Lübeck or Rostock, but I did see Münster some days after you had bombed it several times. There was hardly a whole building standing in the middle of the town. House after house was an empty shell of blackened walls. Street after street was a mere avenue between heaps of rubble.
>
> Münster was, in fact, a sort of Guernica on a larger scale – a terrifying demonstration of what persistent mass bombing could do in a limited area. Increase the scale, and you will get the same result in wider areas. There is a limit to what people can stand, and, as I say, that limit can be coldly calculated and achieved.

On January 1st, 1943, *The Times* reported that 250 acres of this small city had been devastated. More recently, on October 13th, 1943, recording that 'Münster has frequently been bombed', and that 'The territory of the Reich is being battered and laid waste as never before in the history of modern warfare.' Noel Panter, Zurich correspondent of the *Daily Telegraph*, quoted a German spokesman's description of a British and American raid on Sunday, October 10th:

> The attack took place while worshippers filled the churches, and that is one reason for the high casualties. As Münster has no industries worth mentioning, the town and the immediate vicinity were not strongly protected.

Noel Panter goes on to comment:

> whatever industries Münster may or may not possess nowadays, the correspondent carefully conceals the fact that it is a big army supply and administrative centre and an important communications centre.

Having previously remarked on the 'quite unusual sense of reverence' with which German spokesmen now describe Münster as an 'episcopal city' he adds:

> As for the contention that the raid was of an especially ignoble kind because the churches were filled, *even Goebbels could hardly be able to persuade the German people that the British and American Air Force can be expected to consider the times of church services in enemy countries.*

The italics are mine. According to this argument, it was entirely reasonable of the Luftwaffe to bomb to death a number of children attending Sunday School at a church in Torquay.

The description of Münster as 'an episcopal city' is not incorrect, as a Saxon bishopric was founded there by Charlemagne in 805. Since the War, the anti-Nazi pronouncements of its courageous Bishop, Count von Galen, have become well known to those who follow events in Germany. The town possessed numerous medieval buildings, including the Gothic Church of St Lambert (fourteenth century), and a university founded about 1771. In its Town Hall, built in 1335, the Treaty of Westphalia, which concluded the Thirty Years' War, was signed on October 24th, 1648. Of this War, James Harvey Robinson wrote in *The History of Western Europe*, a standard school textbook which was first published in 1902:

> The accounts of the misery and depopulation of Germany caused by the Thirty Years' War are well-nigh incredible. Thousands of villages were wiped out altogether; in some regions the population was reduced by one half, in others to a third, or even less, of what it had been at the opening of the conflict. The flourishing city of Augsburg was left with but 16,000 souls instead of 80,000. The people were fearfully barbarized by privation and suffering and by the atrocities

> of the soldiers of all the various nations. Until the end of the eighteenth century [i.e. for 150 years] Germany was too exhausted and impoverished to make any considerable contribution to the culture of Europe.

The *Daily Telegraph* correspondent, however, boasts that the territory of the Reich is being laid waste as never before in the history of modern warfare; in other words, that the 'atrocities' of its enemies exceed even those of the Thirty Years' War. Inevitably, therefore, the after-effects, in terms of privation and barbarism, will be still graver and more prolonged. Is this a prospect to which even the least thoughtful among the British and American peoples look forward with enthusiasm?

'A British commentator said the other day that there were now 17 German cities in which we had made civilized life impossible,' stated 'Critic' in the *New Statesman and Nation* for December 18th, 1943. He added: 'No doubt he is quite right, but has it occurred to him that civilized life is not the only kind? Men can live like beasts.'

12 Nuremburg

Within recent years, Nuremberg has acquired an unenviable notoriety as the scene of Nazi rallies. It had, however, many centuries of history to its credit before the Nazis were ever heard of, having been made a Free City in 1219. Less insensate ages than our own are likely to regard as catastrophic the fact that the choice of Nuremberg as a Nazi meeting place was used by Bomber Command as an excuse for destroying the heritage of those centuries.

Nelson's Encyclopaedia records of Nuremberg:

> It still retains its ancient walls and moat, and is one of the richest towns on the Continent in medieval buildings and works of art: Albrecht Dürer, Veit Stoss, Peter Vischer and Adam Kraft lived and worked here . . . The churches are full of priceless paintings, statuary and carvings. The Castle, dating from 1050, was enlarged by Frederick I (Barbarossa), and has served as a residence for many German Emperors. Of famous collections the Germanic museum is the most valuable, and a remarkable library, dating from 1445, is preserved in the old Dominican Monastery. Its picture gallery contains masterpieces by Holbein, Dürer and others.

Of this historic city – as priceless to Germans and all students of German culture as is Oxford to ourselves – *The Times* (January 1st,

1943) recorded that 106 acres had been devastated by the end of 1942. In addition to its irreplaceable treasures, Nuremberg had other characteristics of medieval cities, such as narrow streets and ancient ramshackle dwelling houses closely crowded together. In his popular *Berlin Diary*, William Shirer, describing a Nazi rally which he was sent to report, writes of

> the narrow streets that once saw Hans Sachs and the Meistersingers . . . the Gothic beauties of the place, the façades of the old houses, the gabled roofs . . . the streets hardly wider than alleys . . . the beautiful old Rathaus.

On these narrow streets and crowded inflammable houses, the R.A.F. dropped 1,500 tons of bombs on August 10th, 1943, and another 1,500 tons on August 27th. Forty-nine bombers were lost in these raids. The German official figures – almost certainly an understatement – gave the number of the dead as 3,947. On August 29th, 1943, the *Sunday Express* recorded:

> Few towns, even in Germany, can ever have received so shattering a blow in 40 minutes as medieval Nuremberg, the Bavarian 'holy city' of the Nazi Party, which was the target of the vast armada of bombers that roared for more than an hour over south-east England late on Friday.
>
> The result was summed up in one pregnant sentence by a rear gunner on his return. He said: 'I reckon we knocked the whole place flat.'

Under the heading, 'We've Finished the Job Properly', the 'story of this great raid as told to the *Sunday Express* by the men who made it' was given by Edward J. Hart, *Sunday Express* Air Reporter:

> Nuremberg, centre of some of Germany's most vital war industries, was a seething bonfire when our very strong force of four-engined bombers left the scene. Crews returning at dawn brought glowing descriptions of the effects of their heavy bombs and incendiaries.
>
> A solid red core of leaping flames, with columns of jet black smoke billowing up to 15,000 feet and visible 150 miles away, was the word picture painted for me by Flight-Sergeant John Crabb, of Glasgow, navigator of 'S for Sugar', making his twenty-second raid on Germany.
>
> 'I never imagined a town could burn like that,' declared the rear

> gunner of 'A for Apple', Sergeant Harry Smith, a Cardiff man, on his thirty-seventh raid.

Another pilot apparently regarded the whole grim expedition as a colossal joke:

> Lambie seems to have had a thoroughly enjoyable night. 'We had our first bit of fun about ten miles inside France on the way in,' he said . . . 'It reminded you of illuminations in Blackpool – lovely to look at.' Nuremberg, he said, was a good runner-up to Berlin . . . 'With plenty of fire, smoke, flak and searchlights it was everything you could wish for,' he commented. Only one thing marred his joy. In civil life he was a concert singer. Over the target he had promised the crew the 'Prize Song' of *The Meistersingers of Nuremberg*, but just as he started, the intercom system failed. Lambie refuses to believe it was accidental.

In *Das Volksrecht*, Zurich, a Swiss correspondent reported on October 2nd, 1943: 'The whole of Nuremberg is one great ruin, whereas the Siemens-Schuckert works, which were probably the object of the bombardment, received no damage.'[4] Perhaps this is why the *Daily Express* Air Reporter asserted, on January 3rd, 1944, that 'Nuremberg and Munich have yet to be finished off.'

One cannot help wondering whether rear-gunner Sergeant Lambie will feel so elated about the destruction of Nuremberg 20 years hence.

13 Three Prussian Towns (*Anklam, Marienbad, Remscheid*)

Three first-hand reports recently received describe the effect of 'obliteration raids' upon small towns where the area, the population, and hence the capacity for defence is limited, and the power of recovery almost nil. The first two accounts were given to the *News-Chronicle* (October 26th, 1943) by repatriated British prisoners of war:

> With the wounded party came a group of R.A.F. men, some of whom had been imprisoned more than three and a half years. These men had flown Whitleys, Fairey Battles, and other planes whose names are almost forgotten.
>
> These and other pilots passed through Anklam on their way to the port from which they were to be repatriated. Flt. Lt. Howard, a New Zealander, a fighter pilot, told me: 'We were absolutely staggered at the sight. It seemed as if the whole place, works and everything, had been knocked absolutely flat. It was as though it had been smashed over and over again. There was just nothing left.'

Whether the sight of the actual damage done will reduce the enthusiasm of our pilots for bombing Germany has yet to be ascertained. Sergeant Roberts, R.A.M.C., carried on the story:

> We embarked on the train and passed through Marienburg on the way back. It was flat, dead flat; everything the American bombers had set out to smash they had smashed irremediably.
>
> One of the things that struck me as I looked at the German women we saw on stations was their yellowness. They were yellow with undernourishment. Children were the same.

A description of a raid on Remscheid – slightly larger than Reading, with a population of 107,000 – appeared in the *Sunday Express*, August 1st, 1943:

> Remscheid, medieval Rhineland city and centre of Germany's machine tool industry, had its first and probably last R.A.F. raid . . . A navigator said that the outline of the blazing mass below exactly corresponded with the contour of his target map . . . Remscheid measures one and a half miles from east to west, and three and a half miles north to south.

The night attack of July 30–31st on Remscheid was further described by Group Captain Hugh Edwards, V.C., in the *Daily Mail*, October 13th, 1943. The attack, he said, was over in 20 minutes, and

> photographic reconnaissance two days after did not show the town – for the simple reason that the town had ceased to exist . . . The real damage on a big scale is caused when the fires become uncontrollable . . . The aircraft attack . . . is one continuous concentration in order to saturate the defences . . . Crews have no time to dwell on the terrible nature of the attack being carried out down below; they are intent on carrying out their mission and preserving themselves.

Doubtless these crews do not dwell today on the awful cost of that self-preservation to helpless civilians. Doubtless they do not picture the frantic children pinned beneath the burning wreckage, screaming to their trapped mothers for help as those 'uncontrollable fires' come nearer. But what will be the effect of their deeds upon the more sensitive of these young flyers when in future years they come to know what 'the terrible nature of the attack' really meant, and have time to think about it? They may, perhaps, be forgiven by some of their surviving victims, but will they ever forgive themselves? What aftermath of nightmare and breakdown will come? Has any nation the right to make its young men the instruments of such a policy? These are the questions that we ought to be asking ourselves today. Thousands of mothers of young airmen must already be asking them in their hearts.

14 The Ruhr and Neighbourhood

The Ruhr Valley first became famous in recent history when the French occupied it in 1923–24 in an attempt to enforce the astronomical Reparations payments demanded from Germany by the Treaty of Versailles.

The towns of the Ruhr and neighbouring districts – the industrial heart of north-west Germany, covering a total area of approximately the same size as Greater London – have suffered more from our repeated mass raids than any comparable section of Germany. 'Not a town in the Ruhr has escaped being hit,' stated Basil Cardew, *Daily Express* Air Reporter, on January 3rd, 1944:

> The damage there is amazing. Harris has personally spent many hours deciphering the mosaic of the Ruhr reconnaissance pictures that show the holocaust. *He knows of no centre in such an incredible condition anywhere else in the world.* (Italics mine)

Newspaper descriptions of these attacks have made the names of the 'targets' familiar to British readers: Essen, Duisburg, Krefeld, Dortmund, Bochum, Wuppertal, Hagen, and many others. The raids on these towns have been too massive and frequent to describe in detail. Only a few sample facts and extracts can be given.

Bochum

This town has often been attacked, and the official German figures give the number of its dead as nearly 5,000. On June 29th, 1943, the German radio reported that in an R.A.F. raid on June 12th, an orphanage had been bombed and 100 children killed.

Dortmund

In May 1943, a summary of the work of Bomber Command stated that 150 acres of the centre of Dortmund (containing commercial buildings and flats) had been devastated. German official figures of those killed are 15,008.

Duisburg

On May 21st, 1943, opening Norwich 'Wings for Victory' Week, Captain Harold Balfour, Under-Secretary of State for Air, declared: 'We have ticked off on our "city by city" calendar Dortmund and Duisburg.' The number of dead in Duisburg is officially 4,763.

Essen

On July 27th, 1943, Essen, the home of Krupps' great factory, was described by Ronald Walker, the *News-Chronicle* Air Correspondent, as the 'most bombed city'. At this date the destruction of Hamburg was not yet complete. Essen was the second city, after Cologne, to receive one of Bomber Command's new 1,000-bomber 'saturation raids' during the early summer of 1942.

A German letter quoted by *The Listener*, May 13th, 1943, described a heavy R.A.F. raid on Essen on March 5th:

> It was an inferno: bomb followed bomb; streams of phosphorus flowed from above and incendiary bombs fell without interruption. It is a miracle we are still alive. Our district is completely in ruins, and only western parts of Essen remain standing. We are all completely worn out.

On July 3rd, 1943, the *News-Chronicle* stated that 100,000 people in Essen had no roof over their heads. Not content with this measure of suffering and desolation in an armament-making city, Bomber Command delivered another 2,000 tons of bombs in 50 minutes on July 25th:

> By the end of that time smoke from a mass of fires was rising to over 20,000 feet . . . Three times previously 1,000-ton attacks had been made on the city, and by the night of May 7–8th the total tonnage of bombs unloaded on the five square miles of Essen topped 10,000 tons . . . Early last month stories coming out of Germany spoke of Essen as a dead city. (*News-Chronicle*, July 27th, 1943)

Krefeld

On June 23rd, 1943, an R.A.F. pilot, quoted by the *News-Chronicle*, stated: 'If you can imagine a blaze five or six times as big as the Coventry one, you get some idea of what Krefeld looked like last night.'

Wuppertal
According to the *News-Chronicle*:

> more than 1,000 acres of the industrial town of Wuppertal were devastated in Bomber Command's attack on the night of May 28th . . . One thousand acres of devastation in a town of 200,000 inhabitants means that to all intents and purposes that town has disappeared.

The *Daily Mail* added on July 5th, 1943: 'First the R.A.F. took Barmen, the eastern half of the city [i.e. Wuppertal], and "almost wiped it out". Less than a month later, Elberfeld, the western half, had its turn.'

Further grim details of these raids were given by a Swiss correspondent in *Das Volksrecht* on October 2nd, 1943:

> According to the police officer, frightful scenes occurred in Wuppertal as the city is situated in a valley and possessed narrow streets which made any flight impossible. Numerous victims ran round aimlessly like burning torches until they died. Eighteen thousand people were killed by the bombardment.

The Eder and Mohne dams
In May 1943 came the breaking by bombs of the Eder and Mohne dams. No attempt appears to have been made to warn the helpless populations of the flooded valleys, with the result, as recorded in the *Sunday Express* for May 30th, that 'Reuter's special correspondent in Stockholm cabled last night that according to good authority the number of casualties in the Eder and Mohne dams bombings is 70,000.'

Before this, the *News-Chronicle* had commented on May 19th:

> Westphalia has already been bombed on a scale unknown outside Germany. Not even at the height of the blitz against Britain has the misery of our people compared with that of the Ruhr. Now comes a new terror – the devastation of scores of thousands of homes by flood.

The *National News-Letter* for June 24th added further details:

> The explosion of the Mohne dam was catastrophic. It started with a sharp tone which suddenly changed into the rushing and roaring of water which swept everything along with it through the Ruhr districts

> and hills. Many old historical parts of Soest were simply swept away. The water found its way into mines, and hundreds of workers were surprised by the water during the night shift. Many of them were drowned as the way out was completely blocked by the water . . . There was no drinking water available in many areas.

In the *News-Chronicle* of June 11th, an eyewitness quoted by Stockholm reported the scenes in the graveyards where the German people buried their families as being 'indescribable'.

Summary of the Ruhr Damage

On June 6th, 1943, the *News-Chronicle* quoted a Berlin broadcast by a German war-reporter describing raided areas in the west and north-west of the Reich:

> In these areas war gripped the civilian closer than it gripped some soldiers in the front line, and soldiers from the east front passing through stood silent at the train windows bowing before the sacrifice.

'An account of what the Ruhr looks like . . . received by industrialists in France and transmitted to Madrid,' was reported in the *Daily Telegraph* on August 6th, 1943:

> The damage is unbelievable. The Ruhr Valley is only one endless line of debris covered with twisted steel and the smoke from still burning oil. It is estimated that 250,000 Ruhr workers have suffered from forms of shell-shock as a result of the R.A.F. bombing.

The Spectator added, on September 17th:

> Photographs of the blitzed cities, when examined through the red and blue glasses which throw up the picture in three dimensions, show a devastation almost incredible in its extent and completeness.

Still more recent and graphic details were given by the Swiss correspondent who described the scenes at Wuppertal:

> Over 70 per cent of the big western towns have been destroyed. A comparison with the French territories destroyed during the First World War is impossible; destruction in western Germany today is already many times greater than that of the last War in France. It is characteristic that the destruction is greatest in the centre of the cities, whereas certain industrial establishments which the British reported as destroyed showed, as I witnessed myself repeatedly, no damage

> whatever . . . The cities which our train passed presented a frightful sight. Dortmund, Gelsenkirchen, Oberhausen and Duisberg are great heaps of rubble from which ghostly mineshafts protrude. The heaps of ruins are sometimes so enormous that one often wonders whither they must ultimately be transported.
>
> I did not at first have the feeling that there was enormous destruction in Essen, as all the buildings round the station square were intact. Behind them, however, the terror began. The entire centre of the town and the old part are one pile of rubble, from which rise up only a few isolated houses . . . Remscheid and München-Gladbach have also suffered horribly. It is impossible to enumerate all the stricken cities.

Perhaps some economist with the foresight of John Maynard Keynes can calculate the material consequences for Europe of this wholesale blotting-out of Germany's heavy industries and the cities that housed them. Among those who possess imagination, the psychological consequences by which the future of international relations will be determined are already beyond question.

Notes

1 On the same date (November 25th) the *Daily Herald* quoted an official of the Ministry of Economic Warfare as commenting: 'Heavy raids and lack of good news may depress the Germans, but there is little evidence yet of defeatism, opposition to the War, or sabotage.'

2 Daylight raids on Berlin by the American Air Force during March 1944 caused further heavy damage.

3 *Basler Nachrichten*, September 20th, 1943.

4 This appears to contradict the statement made by J. M. Spaight, C.B., C.B.E., in *Bombing Vindicated* (Bles, 1944) that in a raid on Nuremberg on March 8th, 1943, 'In the Siemens electrical works two-thirds of one workshop covering five acres was destroyed.' But factory damage – unlike the destruction of architectural treasures – is quickly remediable.

5
The Bombing of Nazi-Occupied Europe (including Italy)

Those who protest against 'obliteration bombing' as totally contrary in its conception and consequences to the basic values of a civilized society, are apt to be condemned as 'pro-German' – an inevitable cheap accusation in war-time. I propose therefore to consider very briefly the sufferings inflicted by our bombs upon some of the countries included amongst our former Allies, our present 'co-belligerents', and our future protégés.

France

No attempt can be made here to summarize the effects of three and a half years of extensive raids on French territory. A few typical and recent extracts must suffice to show that German civilians are not the only victims of Bomber Command's policy.

On August 4th, 1943, the *National News-Letter* commented:

> Allied bombing appears to have been very effective in areas visited by my informants. For instance, in Neuchatel probably only about 50 houses are left standing. Large areas of Greater Paris are razed to the ground.

The *Sunday Express* stated on September 19th, 1943, that 'About 900 were killed and 1,900 injured in the Allied raids on France on Wednesday night and Thursday, said Vichy radio yesterday.' On October 17th, the German radio reported: 'St Nazaire does not exist any more . . . 95 per cent of the total population of Nantes were affected by the air attacks.'

One of the most grimly suggestive little comments appeared at the foot of a column on the front page of the *Evening Standard* on November 13th, 1943:

> The French town of Modane-Foureaux was completely destroyed in the latest R.A.F. raid on Modane and surrounding districts, says the Swiss newspaper *Tribune de Geneve*, quoted by Reuter.

> The newspaper adds: 'The latest raid lasted nearly two hours. The alarm could not be given in time, because the sirens did not function.'

The *Evening News* further reported on November 26th, 1943: 'The death roll in the Allied raid on Toulon on Wednesday has risen to nearly 400, said the German-controlled French radio today. The number of injured has passed the 600, it said.'

Factories and working-class areas in or near Paris have suffered frequently from our bombing since the fall of France. A recent typical raid was carried out on the last day of 1943, the 'target' being two ball-bearing plants:

> Paris radio said last night that 11 suburbs of the city were hit. 'Considerable damage was done to property', the announcer said, 'and, according to latest figures, casualties total 150 killed and 290 seriously injured.' He added that a great number of people were still buried under debris and that thousands of persons were left homeless. (*News-Chronicle*, January 1st, 1944)

Italy

On July 27th, 1943, the *News-Chronicle* reported:

> Enemy aircraft no longer patrol the Italian sky. Formations of the Allied air forces in daylight search Italy from end to end and bomb at will, meeting no opposition. Not a gun fires. Not an aircraft leaves the sky.

This seems a somewhat peculiar example of British chivalry towards a beaten foe, unable to resist and anxious to surrender. Already an appeal to the Allies to be spared further attacks, because 'Italy is not capable of answering back even if she wanted to,' had been made in an article in the *Giornale d'Italia*, reported by the *Daily Telegraph* on August 13th. This article said:

> For some time your pilots have not cared much for military objectives and the massacre of our inhabited centres has continued with determined coolness. The last eight or ten days shows the devastation done in Naples, Milan and Genoa to the most famous artistic and historical monuments ... They are hitting at the centres of our cities, knowing very well that their bombs will hit the most famous monuments of the world. Have you ever thought what it means? Have you ever thought of the future?

> . . . The war is a terrible immense tragedy. But it is a temporary one. After the war life goes back to normal. Hate ends, the storm is finished. Your tourists, your artists, and all of you and all your future generations will visit our towns. What will they feel when they look for famous artistic and cultural monuments and hear, 'This was destroyed by the British in 1943'?

Milan

Milan, after repeated raiding, was described by a correspondent of the *Evening News* on August 17th, 1943, as 'A city of the dead, in which nine out of every ten houses have been completely destroyed.'

The Milan correspondent of the Swiss newspaper *Die Tat*, quoted by the *Daily Telegraph* on August 17th, 1943, stated: 'The city is dying. Everything is utter chaos. Fires are everywhere. The centre of the city is a heap of rubble . . . Thousands lost their lives in last Thursday's raid and lie buried beneath the ruins of the city.'

According to a special correspondent of the *Daily Telegraph* (August 17th), roads leading from Milan to the Swiss frontier were choked with thousands of nerve-shattered refugees. Hundreds were said to be perishing by the wayside from exhaustion and lack of food. This was the treatment meted out to the workers in the industrial cities of north Italy, who were allies in the fight against Fascism. More recently, 75,000 residents of Milan were reported to be homeless, and not less than 32,000 cases of influenza to have occurred (*News-Chronicle*, December 15th, 1943).

Turin

On July 14th the German radio announced that 12 churches had fallen victim to the R.A.F.'s latest attack on Turin, and that the part of the university left undamaged in previous raids had now been destroyed.

Under the heading 'Turin a Wreck', the *Daily Herald* of November 15th, 1943, stated: 'Three heavy air raids have rendered Turin practically uninhabitable, according to reports reaching Chiasso.'

Naples

Of Naples, already severely damaged before it became a battleground, the German radio commentator, Gunther Weber, gave the following account, quoted in the *Daily Telegraph* of August 13th, 1943:

> I can hardly recognize this city which I last saw five weeks ago. The last air raid devastated the most beautiful quarters.
>
> When you arrive at the city, having in your memory the recollection of incomparable beauty of its outskirts, the impression is terrible. Thousands of corpses are still buried beneath the ruins of houses.
>
> Thousands and thousands of Neapolitans live in the underground stations. Women with babies in arms, old men, sick and mutilated, lie on platforms. You have to fight your way through them to get out.

Genoa

On August 9th, 1943, the *Daily Telegraph*, quoting Rome radio, stated:

> In Genoa the work of clearing debris in the areas that have been worst hit is continuing. Firemen, armed units, and volunteers have helped to bring the town back to normal.
>
> The church of Santo Stefano, which had been hit during the previous raids, has now been destroyed by fire. The churches of Carignano, San Vito, San Tommaso, San Mercillino di San Marco, and San Siro, formerly the Cathedral of Genoa, must be considered lost.
>
> The Carlo Felice Theatre has been gutted. Severe damage was caused to the Palazzo Rosso and to two other famous palaces.
>
> The Hospital of Cagliera and the hospital of Rivarolo have also been hit. In the Via Roma the Palace of the Rinascente has been completely destroyed by fire. The main railway station was also damaged.
>
> There were scenes of 'indescribable panic' in Genoa, according to travellers reaching the Swiss frontier. At the principal station passengers flung themselves from the windows of a moving train and ran for shelter. Many were injured in the crush. Bombs fell before the alert sounded.

The fourth paragraph gives some idea of the havoc created amongst cultural treasures by an attempt to reach one military objective. On December 15th, 1943, the *News-Chronicle* stated that in Turin and Genoa, which had been badly stricken with influenza, hundreds of thousands were living in caves or in the open, in eight degrees of frost.

Rome

In July and August 1943, two heavy daylight raids were made on Rome. A note in *The Spectator* of July 23rd, 1943, indicates THAT

R.A.F. PILOTS ARE FULLY AWARE OF THE DISTINCTION BETWEEN 'PRECISION' AND 'OBLITERATION' TACTICS AND ARE CAPABLE OF CARRYING OUT THE FORMER WHEN THEY CHOOSE: 'Pilots have made intensive studies of the ground plan of Rome, so that they know exactly what to aim at and what to avoid; their action has been that of precision bombing conducted by daylight.'

In spite of this care, heavy civilian casualties occurred during the bombing of Rome's military objectives. The little town of Frascati, a favourite residential area close to Rome, was treated more savagely, since it had been chosen by the Nazis as their headquarters. Once again, the endeavour to destroy a few ring-leaders by indiscriminate bludgeoning led to a heavy massacre of the innocents. On September 19th, 1943, the following extract appeared in the *Sunday Express*:

> The German-controlled Paris Radio, quoting reports from Italy, said yesterday that the town of Frascati, location of the German G.H.Q. near Rome, had been flattened out by Anglo-American bombers. There were more than 5,000 victims.[1]

Bulgaria

Like the German capital, the city of Sofia has recently been the object of 'obliteration raids'. On January 15th, 1944, the Ankara Correspondent of the *Evening Standard* reported:

> Two more bombings like those of Monday will 'wipe Sofia off the map', said travellers who arrived here today from Bulgaria.
>
> The centre of the capital is already almost destroyed, and Sofia has ceased to be as a city.
>
> Thousands were killed, but it is impossible to assess the total number of casualties as bodies are still buried under debris . . .
>
> Bulgarians who saw the bombing estimate that 400 Allied airplanes took part in the attack, which was a surprise as civilians were unaware of the danger.
>
> They did not take to shelter and this is believed to be the reason for the extremely high number of casualties.

On January 19th, the *Daily Herald* quoted reports from Bulgaria stating that over 300,000 people had been evacuated from Sofia.

Austria

This German-occupied country, to which the Moscow Conference promised independence, has lately begun to suffer severely from raids. Towards the end of October 1943, there occurred, on Wiener-Neustadt, one of the now frequent attacks in which cloud obscured the target, thus making impossible any attempt to distinguish military objectives from residential areas. 'The bombing results could not be seen,' reported the B.B.C. on October 25th.

Note

1 Even Florence, one of Europe's loveliest cities, has not escaped. Its marshalling yards were attacked on March 11th. Press reports stated that pilots had conscientiously endeavoured to avoid damage to art treasures, but Rome radio announced 'many casualties'.

6

The Consequences of Obliteration Bombing

What are the effects of our policy of mass bombing upon ourselves? What are they likely to be for Germany, with its civilian casualties already amounting to over a million, and the devastation of its ruined cities far exceeding the damage done to the battle areas of France between 1914 and 1919?

On December 18th, 1943, *The Times*, in an article entitled 'Through German Eyes', quoted a prophecy by the Berlin Correspondent of the Swiss *Neue Zürcher Zeìtung* to the effect that

> a long-term result of the present heavy bombing of German cities will be an utter proletarianization of the German people, and the complete disappearance of what is left of the German middle class. This time, he says, it will be worse for them than the inflation of the 1920s because the actual property of the bourgeoisie is being destroyed.

The *Christian News-Letter* for December 29th, 1943, produced further evidence of the same kind in the following extract from a correspondent who is exceptionally well informed on the European situation:

> It must be added that the wholesale bombardments which involve the complete blotting out of whole cities have the same effect [as totalitarianism]. Men and women who had still a home and a job to defend, have suddenly become people who have nothing to lose and are thus thrown into the mass of uprooted creatures who are merely the passive playthings of forces which they do not comprehend. At the same time these bombardments create the impression that the whole world has gone totalitarian. It is believed that no country recognizes any longer the limits of consideration of human life and of moral standards. It seems that there is nothing left except the war of all against all.
>
> Thus total warfare achieves the work of destruction begun by totalitarianism. The result is a general deadening of the sense of responsibility and of purpose. Life becomes just a matter of survival.

> Everything else becomes indifferent. Any system of government will do as long as it gives bread and security. Moral standards belong to the past world of tranquility and organic relationships. Human life is very cheap, and if one finds that the disappearance of this or that person is necessary for one's safety or prosperity, that person will have to disappear.[1]

Since the liquidation of the middle class in Germany after the last War was a main cause of the rise of Hitler. we may justly ask what new horror of gangsterdom will ultimately arise from the still greater chaos now being created by the R.A.F. amongst Germany's population.

The results for ourselves may be twofold, the one at present speculative, the other certain. Physically, we may suffer costly reprisals in the near future, even though they may come in the form of a desperate blow struck by a beaten enemy in the final agony of the struggle. Morally we are already involved in a process of deterioration which displays itself in a loss of sensitivity, and in words and actions showing callous indifference to suffering.

On June 24th, 1943, the *Daily Express* quoted 'a Ruhr front-line reporter' as saying on the German radio:

> What . . . the people along the Rhine and Ruhr have had to endure is unbelievable.
>
> The whole attention of the German people is focused at present on the inhuman killing of unarmed, defenceless children, women, and old people . . . The German people has been seized by a wave of hatred which has in no way the effect of making their hearts weak.

At a Berlin reception reported in *The Times* on September 28th, 1942, Herr von Ribbentrop said:

> The future will show whether Churchill's bomb warfare against the civilian population is a good or bad idea. Every single bomb, every destroyed home, every dead person makes the German people more determined to make the British pay.

On October 13th, 1943, after 18 months of severe obliteration bombing, Noel Panter admitted in the *Daily Telegraph*: 'Throughout Germany, there is public clamour for reprisal raids on Britain.' A month later, on November 15th, the *Evening News* reported a 'Hate Britain' meeting at Mannheim:

> More than 30,000 people, carrying spades and axes, attended a meeting addressed by Ley, German labour Front chief, at Mannheim before starting to clear away debris from Allied raids, said the German radio today. Paris radio says that the meeting was called 'to display hatred for Great Britain and to protest against the savage bombing of German towns'.

On November 28th, 1943, after Berlin had received, in the words of the *Sunday Express*, a 'third of its "reduction" tonnage', Robert Ley was reported as writing in *Der Angriff*:

> Berlin stands like iron. We shall never capitulate. An eye for an eye and a tooth for a tooth. Defiance and revenge . . . The suffering which Britain in its blind madness has frivolously inflicted on the Germans will not be forgotten.

At the end of December (December 24th), the *Daily Mail* indicated that British airmen captured at Nuremberg might be the first Allied prisoners to be put on trial for 'war crimes' as a reprisal for the Soviet war-guilt trial of German officers at Kharkov.

In a B.B.C. talk on June 30th, 1943, Ellen Wilkinson said: 'The enemy must, in view of our terrible raids, either hit back at us, or admit his own weakness.' One of our own Ministers of 'Home Security' thereby admits that retaliation for our raids may mean a heavy loss of life amongst British people in that process of 'shortening the War' by experimental bombing which is favoured by the Archbishop of York and others.

Reprisal threats from Germany in terms of 'secret weapons' vary from rocket guns based on the Channel Coast, to 'suffocation bombs' which by destroying the oxygen in the air within a radius of 200 yards cause the suffocation of anyone in that area (reported by the Basle correspondent of *La Suisse* and quoted in the *Evening Standard*, December 16th, 1943). In his New Year proclamation to the German people, Hitler himself declared: 'The bombing war against German towns bites into all our hearts . . . Retaliation will come.'

The adoption of the enemy's standards, which our nation earlier deplored, is only one symptom of the moral deterioration and the brutalization of youth, of which this summary contains so many examples, brought about by two years' practice of intensifying cruelty. One of the most deplorable instances of this corruption of values occurred in August 1943, in an article on the bombing of Hamburg in the *Sunday Graphic* (August 29th). Discussing the casualties caused by our raids, the writer – Hilary St. George Saunders, author of *The Battle of Britain* – expresses himself as follows:

> The loss of life has been, if not commensurate with the destruction of buildings and industrial plants, very high indeed.
>
> The figure of 50,000 dead may well be a conservative estimate and there is little reason to doubt the accuracy of a Reuter report that 18,000 of the citizens of Hamburg perished when the tunnel beneath the Elbe received a direct hit.
>
> Is it too much to believe that *a similar success will be repeated* not only in the case of Berlin, but also in those of the 22 other cities of Germany with over 300,000 inhabitants? I do not think so. (Italics mine)

The Pope, speaking to 19 cardinals on his birthday in June 1943, truly stated:

> The progressive use of means of war which make no distinction between military objectives and non-military targets, and the increasing violence of the technique of war, draw attention to the sad and inexorable race between actions and reprisals, which happens to the detriment, not of certain particular peoples, but of the whole community of nations. (*News-Chronicle*, June 3rd, 1943)

A striking example of callousness towards children was embodied in the B.B.C. broadcast to Germany on June 1st, 1942, already quoted on page 104. Other instances of insensibility were produced by some of the 'Wings for Victory' weeks held during 1943, and by exhibitions of air-raid photographs.

At Wincanton, in Somerset, a Children's Competition was organized in connection with the raising of funds to buy Halifax bombers, in which one item ran as follows: 'Class 4. Write a short essay or poem. Subject: The target on which you would like the three Halifaxes to be sent and why.'

Later in the summer, the *Leicester Mercury* recorded:

> Melton's 4,000lb bomb has been delivered on Germany, together with the rude remarks attached to it. The bomb case, it will be remembered, was on view as part of the Melton A.T.C. Exhibition during 'Wings for Victory' Week, and hundreds of Meltonians paid 6d. for the privilege of attaching a 'rude slip' to the case.

On December 31st, 1943, *The Times* reported a similar instance from Canada, which appears as an obscene blasphemy to those who remember the meaning of Christmas:

> Among the bombs dropped [on Berlin on December 29th] by the R.C.A.F. Bomber Group, was a 'block-buster' for which the citizens of Ontario subscribed $1,000,000 in war savings stamps. The bomb bore the names of cities, towns and villages, and the inscription read: 'Ontario citizens' special Christmas card to the Reich.'

A friend, visiting Ford's (Regent Street) Exhibition of Air-Raid Photographs, heard a comment on the pictures of the floods caused by the broken dams: 'Some of the sods were drowned, but not enough of them.'

The National Savings Committee recently issued an advertisement which included these words: 'We heard them go . . . counted them . . . listened to that never-ending roaring stream . . . of machines, carrying our men . . . to tear wide open again and again . . . the bomb-drenched cities of the enemy.'

And to what ultimate end? Will nothing make real to us the abomination of utter desolation which Bomber Command is preparing for post-war Europe? The *National News-Letter* of July 5th, 1943, warned its readers: 'When the clouds of war have passed there will be terrible devastation in Germany – both physical and moral.' *The Spectator* added, on July 30th, 1943:

> We shall not know until we occupy Germany just how much damage our raids have done; for while our photographs told the truth, it is always less than the truth, and what we have repeatedly found when we occupied enemy sites in Africa and Sicily justifies our assuming that the under-statement is considerable.

It is hardly surprising that Mr John Masefield, the Poet Laureate, warned his hearers in a speech at the Annual Luncheon of the National Book Council on October 26th, 1943: 'Europe totters on the brink of a dark time which may conceivably be the darkest time the world has ever known.'

Note

1 In his recently published book *Bombing Vindicated* (Bles, 1944) Mr J. M. Spaight maintains that the bomber has 'saved civilization'

because: (1) Its use avoids the mass slaughter characteristic of such infantry battles as the Somme, 1916, and Passchendaele, 1917; (2) Its cost in *British* civilian casualties was approximately only 90,000 during the blitz period of September 1940 to May 1941. He also uses the curious argument that bombing is to be commended because its effect upon the enemy is 'trivial' compared with that of blockade.

Mr Spaight omits any exact mention of the number of German, Italian, and ex-Allied casualities caused by the Anglo-American air assault. Nor does he point out, save incidentally ('Bomber Command, if it has done nothing else, has proved itself an efficient organizer of mass-migrations'), that the fragile structure of European civilization is threatened by many other factors besides the loss of human life. These include the breakdown of transport, food distribution and other essential services; epidemics due to dirt and disorder; and the existence of millions of dispossessed people with 'nothing to lose' who are leading sub-human lives, and whose gradual barbarization (as in the Thirty Years' War) contains the seeds of anarchy and chaos. But the defensive, apologetic tone of this 'Vindication' suggests considerable uneasiness of conscience despite its protestations.

7
Bombing and Public Opinion

It is difficult to estimate the amount of public misgiving which exists on the subject of 'obliteration bombing'. A valuable article entitled 'Vengeance', by Mass Observation, in the *New Statesman and Nation* of February 12th, 1944, testified both to the divided mind of this country on reprisals, and to the degree of vague discomfort which exists today. 'It was regularly found that, after a blitz, people in bus, street and pub seldom talked of getting their own back,' recorded the writer regarding his experience in 1940, and went on to state that 'nearly one person in four expresses feelings of uneasiness or revulsion' about Britain's present methods of bombing.

Even more difficult is it to find published expressions of concern, which are, of course, officially discouraged in every possible manner. The few letters which appear in the Press – such as that of Mr William Johnstone in *The Spectator* of September 24th, 1943 – are probably quite unrepresentative of the number received. Mr Johnstone's letter inquired:

> Why is it that so many religious leaders, politicians and journalists, who denounced German barbarism during the heavy raids on this country, now either applaud such methods when they are adopted in intensified form by the Allies, or acquiesce by their silence?

It is significant that – as Mass Observation and the organizers of the Gallup Poll found in 1940 – the few protests which do find their way into newspapers appear to come from citizens who, through knowledge of suffering, have outgrown the adolescent vengefulness of immune people living in safe areas. Mr Johnstone – who followed his letter of September 24th with another in February 1944 criticizing Lord Cranborne's reply to the Bishop of Chichester in the House of Lords (see page 172) – wrote from Lambeth, S.E. A similar letter appeared on November 26th, 1943, in the Hull *Daily Mail* from W. S. Robertson, 23 Park Avenue, Hull – a city which suffered severely in the raids and is still intermittently attacked:

> When Hull was mercilessly blitzed [writes this correspondent] and civilians were the chief sufferers, we condemned out of hand such barbarous methods of warfare; and what was wrong for Germany to do then does not become right for us to do now, merely by the passage of time and the fact that we seem to be 'on top'. You can't maintain that it is right for Britain to bomb whole areas indiscriminately without justifying the equally inhuman blitzing of Hull.

Many people, however, accept the Government-inspired pronouncements of the B.B.C. as a substitute for newspaper reading, and those who do read newspapers seldom see more than one. In consequence very few know more than a small percentage of the facts, and still fewer bring imagination to bear upon their meaning. The popular tendency, in this as in all wars, is to erect a psychological barrier against any fact which leads to the discomfort of an uneasy conscience.

Again, the great majority have a quite inadequate knowledge of the immense emergency powers which the Government conferred upon itself in 1940. Their ignorance is combined with large and vague fears regarding the manner in which these powers might be used against the individual who criticizes or protests. Thus, as always, a combination of ignorance and timidity keeps many silent who would like to speak. Public disquiet expresses itself mainly through half-surreptitious conversations in railway carriages, shops and restaurants, carried out with a nervous eye on posters depicting the dire consequences of 'Careless Talk'. It is, however, noteworthy that the general acclamation of the R.A.F. pilots who defended this island in the Battle of Britain has not been conspicuously transferred to their death-dealing colleagues in Bomber Command; and that the B.B.C. itself often comes in for criticism. 'A few days ago', ran a letter written from Pietersburg, South Africa, on November 23rd, 1943,

> the B.B.C. told us that 12,000 people had been rendered homeless by a raid over a German town. . . . But the people I have spoken to didn't seem so enthusiastic as the B.B.C. seemed to expect; perhaps if you've ever been rendered homeless, as many of our fathers and mothers were in the Boer War and as so many English people have been in this war, the fact that thousands of others are in the same plight is less heartening than the unimaginative suppose.

The past few months, at any rate, have been marked by some courageous pronouncements from leaders of opinion, among them one or two well-known Churchmen. In July, 1943, Dean Inge, writing in the *Church of England Newspaper*, stated: 'I believe when the war is over we shall be very sorry for what we have done. European civilization is one and indivisible.' On December 7th, 1943, the *Daily Mail* reported the Bishop of Gloucester as saying of our policy of bombing Germany: 'It is a form of warfare I do not like. It is very largely a war against civilians. It is a barbaric war. It means the destruction of all the fruits of civilization.'

Writing in the *Evening Standard* of January 4th, 1944, on man's unlimited propensity for destruction, Major-General J. F. C. Fuller said:

> In the last War it was the artillery battle; in this war it is air bombardment. By means of the one he obliterated entire battlefields, and by doing so denied to himself all possibility of exploiting the initial success gained by becoming bogged in the slough he created. By means of the other he has annihilated great cities and vast industrial areas, and *in consequence has pulverized the very foundations upon which eventual peace must one day be built*. (Italics mine)

Two days later, in the *Daily Mail*, Mr Bernard Shaw made a pronouncement of characteristic vigour:

> As to atrocities, we have rained 200,000 tons of bombs on German cities; and some of the biggest of them have no doubt fallen into infant schools or lying-in hospitals. When it was proposed to rule this method of warfare out, it was we who objected and refused. Can we contend that the worst acts of the Nazis whom our Russian allies have just hanged were more horrible than the bursting of a bomb as big as a London pillar-box in a nursery in Berlin or Bremen? . . . German papers, please copy. Our enemies had better know that we have not all lost our heads, and that some of us will know how to clean our slate before we face an impartial international court.

On February 9th, 1944, the Bishop of Chichester (whose record on bombing, as on famine relief, has been one of consistent courage) raised the question of British bombing policy in the House of Lords, and was supported by Archbishop Lang (recorded in the next section, 'Comment in Parliament'). Today there are probably few who would disagree with the solemn appeals and warnings addressed to all belligerents by the International Red Cross Committee on July 24th, 1943, and again on December 30th, 1943,

> urging them again to safeguard the natural claim of the individual to justice and protection from arbitrary measures, to refrain from holding him responsible for acts not committed by him, and to renounce unwarranted destruction, and, in particular, the use of harmful methods of warfare which are prohibited by international law.

'The International Committee observes', adds the second statement,

> with great and increasing disquiet the steady aggravation of methods of warfare affecting civilian lives and property devoid of military importance but the destruction of which is an irreparable loss to civilization. The principle laid down by international law, according to which the legitimate destruction of military forces and objectives may not expose lives or property of a non-military character to hurt or risk out of all proportion to the importance of the aim in view, seems to be relegated now more and more to the background in favour of the unreserved pursuit of total warfare.

8
Comment in Parliament

Expressions of opinion adverse to 'obliteration bombing' have been fairly frequent in the House of Commons, since M.P.s have greater privileges, more knowledge, and hence less fear, than the general public. Even so, protest has usually been left to a handful of courageous Members, prepared to risk unpopularity in the cause of maintaining civilized standards amid a blind drift into barbarism.

On March 14th, 1943, *The Sunday Times* reported as follows:

> Our power continues to grow, and, to the lasting credit of the House of Commons, Members made it plain that they wanted a reassurance that the original clear distinction between military industrial objectives and the indiscriminate dropping of as much high explosive as possible on congested areas was not being progressively abandoned, just as our superiority should remove any temptation to lower our standards. Captain Balfour gave the reassurance, but the Air Ministry can hardly make it too plain that we do not take our standards in these matters from the enemy; for we are fighting for the preservation of a civilization and he is not.

This demand for reassurance occurred 12 months ago. Is the House better satisfied today that 'we do not take our standards in these matters from the enemy'? Does it feel that the methods of Bomber Command are likely to result in 'the preservation of a civilization'?

A fortnight later, Mr R. R. Stokes asked the Secretary of State for Air whether on any occasion instructions had been given to British airmen to engage in area bombing rather than to limit their attention to purely military targets. Sir A. Sinclair replied: 'The targets of Bomber Command are always military, but night bombing of military objectives necessarily involves bombing the area in which they are situated' (House of Commons, March 31st, 1943).

On May 27th, another M.P. asked whether the Government would give consideration to any representations made by the Christian Churches on this issue. Mr Attlee, Deputy Prime Minister, replied: 'We have not received any. We must wait until we do.'

Commenting on this reply in a letter to *The Friend* of June 25th, 1943, Corder Catchpool, a leading member of the Society of Friends, stated: 'This constitutes a direct challenge, almost an invitation, from the Government to the Churches to speak their mind.' But except for an episcopal minority, the Churches have been silent. The Archbishop of Canterbury was too busy in July 1943 to receive a group of clergy and others who were anxious to discuss the implications of mass bombing. Occasional clerical pronouncements strike a definitely anti-Christian note. On July 4th, for example, a letter from Canon Hannay appeared in the *Sunday Express* deprecating German 'squealing' over the bombing of Cologne Cathedral.

On July 28th, 1943, Mr Reginald Sorensen, M.P., asked the Secretary of State for Air 'whether the same principles of discrimination that are applied to Rome are being and will be applied to other cities?' Sir A. Sinclair replied:

> The same principles are applied to all centres. We must bomb important military objectives. We must not be prevented from bombing important military objectives because beautiful or ancient buildings are near them.

In other words, many irreplaceable churches, monuments and other treasures must be destroyed on the off-chance of hitting one railway station or an isolated factory.

Following the series of November raids on Berlin, Mr R. R. Stokes put further questions to the Secretary of State for Air on December 1st, after Sir Archibald Sinclair had informed Wing Commander Hulbert that Bomber Command had dropped approximately 13,000 tons of bombs on Germany during the previous month. The following interchanges then took place, as reported by that day's *Evening Standard*:

> When Mr Stokes asked for the area, in square miles, in Berlin within which it was estimated that 100 per cent of the 350 block-buster bombs recently dropped in a single raid would fall, Sir Archibald said he could not reply without giving useful information to the enemy. Mr Stokes: Would not the proper answer be that the Government dare not give this information? Sir Archibald: No. Berlin is the centre of 12 strategic railways. It is the second largest inland port in Europe. It is connected with the whole canal system in Germany. In that city are A.E.G., the Rhein Metall, Siemens, Focke Wulf, Heinkel, and Dornier factories. (Cheers.) If I were allowed to choose only one

> target in Germany it would be Berlin. (Cheers.) Mr Stokes: Do you not admit that the Government are now approving the indiscriminate bombing of Germany? Sir Archibald: You are incorrigible. I have mentioned a series of vitally important military targets. Mr Shinwell (Soc. Seaham): Will you appreciate that much as we deplore the loss of civilian life anywhere we wish to encourage and applaud the efforts of your Ministry in trying to bring the war to a speedy conclusion? (Cheers.) Mr Simmonds (Con. Duddeston): Are not these bombings likely to reduce, vastly, our military casualties when we invade Europe? Sir Archibald: Yes. Mr Stokes also asked whether the policy of limiting objectives of Bomber Command to targets of military importance had or had not been changed to the bombing of towns and wide areas in which military targets are situated. Sir Archibald: There has been no change of policy.

Hansard's report of the same debate continues:

> Mr Stokes: May I say that the reply of my Right Hon. Friend does not answer this question. Am I to understand that the policy has changed, and that new objectives of Bomber Command are not specific military targets but large areas, and would it be true to say that probably the minimum area of a target now is 16 square miles? Sir Archibald Sinclair: My Hon. Friend cannot have listened to my answer. I said there had been no change of policy.

If Sir Archibald is indeed correct in asserting that the policy of Bomber Command is unchanged, the 'obliteration bombing' of cities must have been the deliberate intention of the present Government for a very long period.[1]

The most recent important Parliamentary interchange on the bombing of Germany occurred on February 9th, 1944, in the House of Lords, when the Bishop of Chichester asked whether, without detriment to the public interest, the Government could make a statement as to their policy regarding the bombing of towns in enemy countries. Extracts from his speech on this occasion are taken from *The Times* Parliamentary Report of February 10th:

> He was not forgetting the Luftwaffe or its tremendous bombing of Belgrade, Warsaw, Rotterdam, London, Portsmouth, Coventry, Plymouth, Canterbury, and many other places of military, industrial, and cultural importance. Hitler was a barbarian. There was not a decent person on the Allied side who was likely to suggest we should make him our pattern . . . The question with which he was concerned

was this: Did the Government understand the full force of what our aerial bombardment was doing and what it was now destroying? Was it alive not only to the vastness of the material damage, much of which is irreparable, but also to the harvest it was laying up for the future relationships of the peoples of Europe? He recognized the legitimacy of concentrated attack on industrial and military objectives . . . He fully realized that any attacks on centres of war industry and transport inevitably carried with it the killing of civilians. But there must be a fair balance between the means employed and the purpose achieved. To obliterate a whole town because certain portions contained military and industrial objectives was to reject the balance.

He would instance Hamburg, with a population of between one and two millions. It contained targets of first-class military importance. It also happened to be the most democratic town in Germany, where the anti-Nazi opposition was strongest . . .

Berlin, the capital of the Reich, was four times the size of Hamburg. The military, industrial, and war-making establishments in Berlin were a fair target, but up to date half the city had been destroyed and it was said that 74,000 persons had been killed and 3,000,000 were already homeless. The policy was obliteration, openly acknowledged, and that was not a justifiable act of war. Berlin was one of the greatest centres of art galleries in the world. It had one of the best picture galleries in Europe, comparable to the National Gallery, and had one of Europe's finest libraries. All these non-industrial, non-military buildings were grouped together, and the whole of the area is reported to have been demolished. These works of art and libraries would be wanted for the re-education of the Germans after the war . . .

There were old German towns away from the great centres which almost certainly would be subjected to the raids of Bomber Command. In these places the historic and beautiful centres were well preserved and the industrial and military establishments were on the outskirts. We had destroyed much; we ought to think once, twice, three times, before we destroyed the rest . . .

He emphasized particularly the danger outside Germany, to Rome. The principle was the same, but the destruction of the main Roman monuments would create such hatred that the misery would survive when all political and military advantages that might have accrued had long worn off. The history of Rome was our own history. Its destruction would rankle in the memory of every good European as does the destruction of Rome by the Goths or the sack of Rome. The blame must not fall on those who were professing to create a better world. It would be the sort of crime which even in the political field would turn against the perpetrator.

> It had been said that area bombing was definitely designed to diminish the sacrifice of British lives and to shorten the war. Everybody wishes with all his heart that those two objects could be achieved, but to justify methods inhumane in this way smacked of the Nazi philosophy of 'might is right'. At any rate to say that it would reduce sacrifice was pure speculation.
>
> 'If there was one thing absolutely sure it was that the combination of the policy of obliteration with the policy of complete negation as to the future of a Germany which had got free from Hitler was bound to prolong the war and make the period after the war more miserable.'
>
> This speech was more briefly supported by Archbishop Lord Lang, who said that the recent attacks on cities like Hamburg, Frankfurt, and Berlin, seemed to him to go a long way beyond what had hitherto been the declared policy of the Government. He emphasized the moral deterioration which this policy was causing amongst our people, and the danger to our standards which its continuation would represent.

Lord Cranborne, in reply, justified the present methods on the usual grounds of military necessity. 'These great war industries', he argued, 'can only be paralysed by bringing the whole life of the cities in which they are situated to a standstill.' He further declared that the purpose of the present air offensive was to achieve the end of the war at the earliest possible moment.

'Such methods', wrote Mr William Johnstone in his letter to *The Spectator*,

> have dangerously lowered the moral Plimsoll line in warfare, and the arguments Lord Cranborne used can be made to justify, if need be, the abandonment of all restraints on grounds of military necessity and, indeed, of 'humanity'. This has a familiar Teutonic ring.

On October 28th, 1943, a somewhat different exchange of opinion had occurred in Parliament.

Mr R. Purbrick (Cons. Walton) asked the Home Secretary whether his attention had been drawn to an Association called the Bombing Restriction Committee, and as their activities are clearly evidence of their pro-German sympathies, will he take steps to restrict such activities and intern the leaders?

The Home Secretary replied:

> I am aware of the existence of the Committee and I have seen specimens of the leaflets issued by them. The scope of propaganda is very limited and its influence on public opinion is negligible. I have no evidence that their sympathies are pro-German, nor have I any reason to question the sincerity of their motives, which they conceive to be purely humanitarian, and I should not feel justified, as things are, in using emergency powers to prevent them from giving expression to their misguided views.

Mr Purbrick continued:

> Is it not evident that if the policy of not bombing any civilian in Germany was carried out we should not be able to carry through our policy of bombing their war production plant, and therefore our war effort would be absolutely nullified?

The Home Secretary said:

> I quite follow the argument of my Hon. Friend, but in my judgement if people sincerely hold the view that bombing should be restricted or abolished, I cannot see that it is terrible to say so. There is no danger that the bombing will leave off, anyway. I would draw the attention of the House to how often I am condemned for using exceptional powers, and how much more often I am pressed to use them.

Mr Rhys Davies concluded the discussion with a supplementary question: 'Is the Right Hon. gentleman aware that the object these people have in mind is to restrict and prevent the bombing of civilians, which is another problem entirely?'

In view of the Fascist power to suppress inconvenient minorities with which he was endowed by the panic-stricken Parliament of 1940, Mr Morrison's attitude is to his credit. One wonders how far he really considers it 'misguided' to deplore the slaughter of the invalid and the aged, and to abominate the brutal massacre of young children cherished by the founder of the faith officially accepted by this Christian country.

Posterity, at any rate, will have the final decision as to which group was 'misguided' and which perceived the truth. Our politicians, who so blandly rationalize the perpetration of atrocious deeds, have still the future to face. The war cannot continue for ever. If we were unable – as a *Sunday Times* writer asserted in 1940 – to forgive Germany for 'bombing the past out of our lives', how can we expect the Germans – who are losing not only their past, but their present and future – to forgive us and the Americans when the day of reconstruction comes? And how can we, whose newspapers hold even the German children responsible for the domination of Hitler, forgive ourselves if we fail to convey our opinions to our political leaders?

The relevance of these enquiries is not abolished by conferring opprobrious names upon those who make them.

Note

1 A book entitled *The Coming Battle of Germany*, by an American author, William B. Ziff, published by Hamish Hamilton in 1942, gives some substance to this belief.

9
Some Suggested Remedies

There are many methods by which the advice of the International Red Cross to renounce unwarranted destruction might be taken. The difficulty, at the stage of spiritual demoralization which this country has reached, is not to discover methods, but to create the will to use them among those responsible. An alternative is, of course, to demand the replacement of these men and women by others more imaginative and far-sighted. Even in a democracy impaired by war, a strong public opinion could still make such a demand effective.

Remedies suggested up to date include open towns, sanctuary areas, the specific proposals of the Spanish Government, and an agreed return by both sides to 'precision' bombing.

On August 10th, 1943, the *News-Chronicle* declared: 'Air warfare has virtually destroyed the convention by which a city may be declared to be open.' The Nazis, however, though in complete control of the air, respected the declaration of the French Government and treated Paris as an open city in 1940. Do we wish to acknowledge our inability to reach even their standards?

The Lord Chancellor (Lord Simon) stated on August 29th, 1943:

> I am not aware that there is any rule of international law which confers immunity from attack upon a town which makes such a declaration. Some people seem to attribute to it as magical an effect as the incantation of 'Open Sesame' in the *Arabian Nights*. The question whether a town can be legitimately bombarded from the air depends not on whether it calls itself 'open' or 'undefended', but on whether it contains any military objectives.

But he added on the same occasion: 'Civilians who are taking no part and are rendering no assistance in carrying on the war should not be made the object of deliberate and direct attack.'

In a memorandum drawn up by the Bombing Restriction Committee, the following proposal was recently put forward:

> Humanitarian considerations demand the recognition by the belligerents of sanctuary areas to which women and aged people could

> be evacuated from all towns having any kind of military objective *in advance of bombing*. In such sanctuary areas non-combatants could live free from the oppression of fear – fear for their own lives and for the safety and well-being of their children. Such sanctuary areas would be especially beneficial to invalids and people of highly nervous temperament who suffer agonies of apprehension if they have not the financial means to travel a long distance to get away from the threatened area.
>
> The sanctuary areas should be located, if possible, within 190 to 150 kilometres of the Ruhr and of the great industrial cities in other parts. It would be an advantage to include towns which have no military establishments or munition works and are situated on unimportant railway lines which do not carry military traffic; such towns, for instance, as Bonn, Homberg, Baden, Heidelberg, and other university towns and health resorts. In Italy there would be no difficulty in finding non-industrial towns to act as the centres of sanctuary areas.
>
> To assure the complete absence of military preparations and personnel from the sanctuary areas, a corps of observers composed of the nationals of neutral countries could be formed. It could be placed under the control of the International Red Cross, or of a neutral commission on which the Vatican would be represented. The Spanish Civil War provides an example of a corps of neutral observers, working on the whole satisfactorily.

Details were added, giving suggestions by which such sanctuary areas should be made recognizable from the air.

The attempt of the Spanish proposal to mitigate the worst horrors of aerial bombing was received here with contempt, on the ground that General Franco's government had no title to express a sudden repugnance towards this form of attack. Without discussing whether Axis sympathies necessarily disqualify Spain from attempting to save at least the lives of innocent German children, it is worth noting that the Spanish suggestions did not meet with the same wholly negative reception in the United States. On July 17th, 1943, Mr Oswald Garrison Villard, former editor of the New York *Nation*, wrote in his American News-Letter:

> To the regret of many whose feelings are deeply stirred by the destruction of innocent men, women and children in the bombings now taking place all over the globe, the Spanish Government's proposal of June 5th of a four-point plan to humanize aerial bombing, has met with no response here. In making the proposal the Spanish

Government insisted that Spain was acting for itself and not for any other nation, and declared that General Franco had already sent a special message to the United States along this line through Myron C. Taylor, the American special envoy to the Vatican. Spain further pointed out that both sides had received unfavourably previous proposals for the limitation of bombing. The four proposals were that the belligerents themselves define the zones in enemy territory which in their opinion constitute objectives for total bombardment for military reasons; that other zones be declared 'partial objectives' and be divided into 'bombardable', 'dangerous' and 'non-bombardable' zones.

Next Spain called for a neutral committee to watch the zones where bombarding is not permitted and to regulate terms of the agreement, and finally that regions in which there are no military objectives, or wherein such objectives are insignificant, be declared non-bombardable zones. Undoubtedly the growing feeling that the United Nations are rapidly bombing Italy and Germany out of the war has had much to do with the indifference to the Spanish proposal. But the news that in the brief hour the R.A.F. destroyed 4,000 lives in the city of Wuppertal, and injured 18,500 more, and has so destroyed the city that it cannot be rebuilt, tries the soul of some people. Few as yet agree, however, with Dean Inge that the United Nations will have cause to look back with deep regret upon the damage done to Europe from the air and the cultural relics destroyed. Popular feeling is still that it is more humane to carry on the policy of complete destruction in order to compel a complete surrender of Germany.

A letter sent on December 31st, 1943, to the War Cabinet, under the signature of Mr J. D. Beresford and other well-known persons, contained the following constructive proposals, 'in order that this policy, by which the British nation will be judged in years to come, may have the free and considered verdict of the British people pronounced upon it':

(1) A full debate on the subject by Parliament within a reasonable time.
(2) The withdrawal of all existing instructions which might lead to biased official propaganda based upon it, pending the holding of the above debate.
(3) The suspension in the meantime of 'obliteration' tactics, or, at the least, the issue of a pronouncement, to the British people and to the world at large, that the sole responsibility for this

> policy at present rests with the War Cabinet, and not with the people.

The suggestion 'that an agreed return should be made to precision bombing' is apt to meet with opposition not only from the total or partial supporters of Sir Arthur Harris' policy, but from those unbending pacifists who maintain that 'humanizing war' involves the acceptance of the war method as tolerable, and that no compromise is possible with those who make use of it.

From the story of our bombing during the past 18 months, only a mental or moral lunatic could fail to draw the conclusion that modern war and modern civilization are utterly incompatible, and that one or the other must go. But the refusal to do anything at all until we can achieve the millennium is too often an attempt to justify doing nothing. In 1625, when Hugo Grotius, reacting against the cruelties of the Thirty Years' War, wrote the book *De Jure Belli ac Pacis*, which laid the foundations of international law, he thereby showed himself one of the first to realize that 'humanizing war' is not an alternative to abolishing it. It is a step nearer to the creation of that state of mind in which the abolition of war will become possible.[1]

In a recent article entitled 'The Invocation of Anarchy', Professor L. W. Grensted, Nolloth Professor of the Philosophy of the Christian Religion in the University of Oxford, writes of 'that self-contradictory humanizing of war which is one of the channels by which its energies are ultimately transmuted, and through which alone it can be associated with creative values'. Every advance along the road to civilization is a compromise between those who desire to go the whole way, and those who do not wish to move at all.

As a nurse in France during the last War, I myself had to care for ten of the first mustard gas cases that came down from the Battle of Cambrai in 1917. Remembering the sufferings of those gassed men and the small percentage of recoveries in the hospital, I for one am thankful for the development of public opinion which has caused the belligerent nations to observe up to date the Poison Gas Convention of 1925, and to deny themselves the dreadful fruit of their scientific researches in this field, even though war itself continues.

But we have no reason to be complacent for refraining, like the Nazis, from this type of chemical warfare. The tortures to which we have subjected civilians, including children, in our 'saturation raids' far exceed the sufferings caused by poison gas between 1914 and 1918. I venture to prophesy with complete confidence that the callous cruelty which has caused us to destroy innocent human life in Europe's most crowded cities, and the vandalism which has obliterated historic treasures in some of her loveliest, WILL APPEAR TO FUTURE CIVILIZATION AS AN EXTREME FORM OF CRIMINAL LUNACY WITH WHICH OUR POLITICAL AND MILITARY LEADERS DELIBERATELY ALLOWED THEMSELVES TO BECOME AFFLICTED.

Note

1 Professor A. L. Goodhart (*What Acts of War are Justifiable*, Clarendon Press, Oxford) states that

> the history of laws of war goes back to the latter part of the Middle Ages, when the influence of Christianity and of chivalry combined to restrict the cruelty of war. The Thirty Years' War was a temporary setback, but the horror which the unrestrained brutality of the soldiers, especially at the siege of Magdeburg, caused throughout Europe, brought about a new development. Hugo Grotius, in his celebrated work *De Jure Belli et Pacis* (1625), did much to advance this by his attempt to state the general principles in concrete form. Further progress was made during the eighteenth century, with the result that the unrestrained cruelty of former times was in large part absent from the Napoleonic Wars. It was, however, after 1850 that the most striking advance was made by means of various treaties and conventions, in which the rules relating to warfare were partially formulated.

10

Conclusion

We can do no less than seek an answer to each new excursion into the dark abyss of inhuman barbarity, for as we become more sensitive and intelligent creatures, our capacity for good and evil alike increases with our knowledge.

It may take centuries yet to abolish war altogether, but within those centuries the terrible refinement of scientific inventions may first abolish man unless he deliberately restrains himself from employing them. One fact, at any rate, emerges from the story of our own mass bombing, by which our leaders have surpassed even the savageries of the Thirty Years' War, and have brought about a still more disastrous setback to the influence of Christianity and of chivalry. If the nations of the world cannot agree, when peace returns, to refrain from the use of the bombing aeroplane as they have refrained from using poison gas, then mankind itself deserves to perish from the epidemic of moral insanity which today afflicts our civilization.

But there is no need to wait for the end of the War before we consider exactly what we are doing, and decide whether we desire the Government which we elected to continue a policy of murder and massacre in the name of the British people. It is now too late to save many of Europe's finest cities, to restore historic treasures reduced to rubble, or to bring back to life more than a million German and ex-Allied civilians. But even though we knew that the rest of the Continent must fall victim to the vandalism of our politicians, the obligation would still lie upon us who repudiate their criminal policy to assert, loudly and clearly, that their deeds are not done by our will or with our consent.

Eight years after the South African War ended, General Botha told Emily Hobhouse that three words, 'Methods of Barbarism' – applied by Sir Henry Campbell-Bannerman to the British use of concentration camps – had made peace and union in South Africa. Today England is less fortunate in the quality of her statesmen; we have no Campbell-Bannerman to raise his powerful voice against evil committed by ourselves. The duty of protest remains none the less; and may still carry its word of hope and healing to an age beyond our night.

Lament for Cologne

You stood so proudly on the flowing Rhine,
Your history mankind's, your climbing spires
Crowned with the living light that man desires
To gild his path from bestial to divine.
Today, consumed by war's unpitying fires,
You lie in ruins, weeping for your dead,
Your shattered monuments the funeral pyres
Of humble men whose days and dreams are fled.

Perhaps, when passions die and slaughters cease,
The mothers on whose homes destruction fell,
Who wailing sought their children through the hell
Of London, Warsaw, Rotterdam, Belgrade,
Will seek Cologne's sad women, unafraid,
And cry: 'God's cause is ours. Let there be peace!'

V.B., The Friend, *June 19th, 1942*

Notes on Contributors

Vera Brittain (1893–1970) grew up in provincial comfort in Macclesfield and Buxton. In 1914, as the First World War was breaking out, she won an exhibition to Somerville College, Oxford, but a year later abandoned her studies to enlist as a VAD nurse. She served in London, Malta, and close to the Front, in France. She also suffered the deaths in the war of her closest male contemporaries – her fiancé, her brother, and two other friends – and the story of her war experiences formed the subject of her most famous book, *Testament of Youth*, which became an international bestseller on its publication in 1933.

Vera Brittain devoted much of her energy to the causes of peace and feminism. She became a pacifist in 1937 when she joined the Peace Pledge Union, and was a leading figure in the peace movement during the Second World War, attacking the Allies' policy of the saturation bombing of German cities.

Vera Brittain was an indefatigable journalist and lecturer, and published 29 books, including autobiography, biography, fiction, and poetry.

Shirley Williams, the daughter of Vera Brittain, began her career as a journalist, having graduated from Somerville College, Oxford. A Cabinet Minister in the Labour Government from 1974 to 1979, she was co-founder in 1981 of the SDP, President of the Party from 1982 to 1988 and Leader of the Liberal Democrats in the Lords from 2001 until November 2004. She is Professor Emeritus at the John F. Kennedy School of Government, Harvard University, and is on the Advisory Council of the Institute for Public Policy Research and the Council on Foreign Relations, New York. She is Co-president of the Royal Institute of International Affairs. Her book, *God and Caesar: Personal Reflections on Politics and Religion*, is published by Continuum.

Aleksandra Bennett, who has focused on Vera Brittain and on British pacifism during the Second World War in her work, teaches Modern History at Carleton University, Ottawa, Canada.